David Vizard's

HOW TO BUILD

Max-Performance

CHEVY

SMALL-BLOCKS

On A Budget

David Vizard

CarTech®

CarTech®

CarTech®, Inc.
6118 Main Street
North Branch, MN 55056
Phone: 651-277-1200 or 800-551-4754
Fax: 651-277-1203
www.cartechbooks.com

Edit by Paul Johnson & Josh Brown
Layout by Chris Fayers

ISBN 978-1-932494-84-6
Item No. SA57

Library of Congress Cataloging-in-Publication Data

Vizard, David.
  How to build max-performance Chevy small-blocks on a budget / by David Vizard.
    p. cm.
  ISBN 978-1-932494-84-6
  1. Chevrolet automobile—Motors—Modification. 2. Chevrolet automobile—Motors—Design and construction. I. Title.
  TL210.V535 2009
  629.25'040288—dc22
                      2008038084

Written, edited, and designed in the U.S.A.
Printed in China
20

Back Cover Photos

Upper Left:
Here is the Scat rotating assembly for a moderate-budget drag race 350 I built. It came balanced as a kit. In its track-ready form the engine made just a couple horsepower shy of 600.

Upper Right:
With early heads (pre-1973) this is the style of chamber that makes power. When milling these heads for compression be sure the groove (indicated by arrow), which circulates water around the plug boss, is still deep enough to do so.

Middle Left:
AED custom-built carbs, like this one for my big-block Chevy, can be expected to run well with your combination right out of the box, yet cost no more than regular carbs bought at your local speed shop.

Middle Right:
Two things to note about this Spintron setup: The bee-hive spring is much smaller than the conventional spring it replaces, and self-aligning rockers ensure rocker alignment and eliminate the need for guide plates.

Bottom Left:
The lobe on the left of this cam is nearly non-existent.

Bottom Right:
The simplest and most cost effective ignition system available for the small-block Chevy is still the HEI. It needs a little hopping up to get the job done, and a custom-built Performance Distributors unit takes care of that.

OVERSEAS DISTRIBUTION BY:

Europe
PGUK
63 Hatton Garden
London EC1N 8LE, England
Phone: 020 7061 1980 • Fax: 020 7242 3725
www.pguk.co.uk

Australia
Renniks Publications Ltd.
3/37-39 Green Street
Banksmeadow, NSW 2109, Australia
Phone: 2 9695 7055 • Fax: 2 9695 7355
www.renniks.com

Canada
Login Canada
300 Saulteaux Crescent
Winnipeg, MB, R3wJ-3T2 Canada
Phone: 800 665 1148 • Fax: 800 665 0103

TABLE OF CONTENTS

ACKNOWLEDGMENTS

To produce any book like this means looking back over one's life and putting some thought into the journey from early childhood to where one is at now. Now, you may ask, how's that relevant here, where the author traditionally thanks all those who helped out on it's production? It's simple: Everyone who has in some way made a contribution to my life has in some way helped toward the production of this book—my thirtieth effort in this direction—and thanking everyone by name is simply impossible. If you have made a positive contribution to my life that has helped me get to this stage, and I do not say this flippantly or lightly, then, thank you.

The easy part of any credits section is thanking the family and friends who surround you on a more or less daily basis. First thanks to my wife, Josephine, and my two girls, Danielle (23) and Jacque (12), for being patient with me and letting me work when family affairs should have figured more prominently. Also, thanks to my many friends whom I would like to mention. To paraphrase a legal document, these include, but are not limited to, Denny, Carl, Don, Bob, Jim, Mike, Mervyn, David, Roger, Lloyd, Terry, Al, Kenny, etc. Please don't feel slighted if you felt you have just fallen into the "etc." category. As I write this, the very last passage for the book, I feel mentally obliterated by running what must be close to two full-time jobs and working on this book each day for over six months.

It seems each book I do has one new person who makes the production of that book, at that moment in time, just that much better. This time around that person was ever-resourceful UNCC engineering student Dusty Kennett. Thanks, Dusty.

INTRODUCTION

The phrase "speed costs money, how fast can you afford to go?" is as true now as ever but as of 2009 your dollar will buy you more speed than almost anyone could have foreseen just a decade ago. Here's a statement of my own, less prosaic, though no less factual. "Hard-core high tech know-how builds speed and cuts cost." I know this for a fact. I've been doing it for 50 years. Let's look at that "hard-core, high-tech" statement in a little more detail. Firstly, "Hard core." This means down to earth, no superfluous frills, bells or whistles, and definitely no fancy "advertising" technology word play. For sure, certain high-performance parts look good on a magazine page but the question is, do the parts concerned increase performance and, more to the point, are they cost effective? Wherever you see the term "hard core" in this book it refers to parts whose design is based on proven engineering principles.

Let's move on to the term "high tech." Here's a really overworked phrase. What some regard as high tech is often aerospace engineering that has been around for ten or more years. The design of advanced aircraft and Formula 1 race cars is front-line high tech. In the more general automotive-performance world, such groundbreaking tech capability exists almost exclusively in the ranks of professional race-car builders. Within these pages it is not my intent to weigh you down with highly technical information then leave you to sink or swim as the case may be. In fact, all will be to the contrary. I'm going to do most of your homework for you. The plan is to relate my experiences then to supply you with the means and the know-how to build engines now that would normally take the accumulation of years of experience to achieve. After reading this book the engines you build will be far more successful than the cost of them would imply. I guarantee it.

Myths and Mistakes

Millions of dollars sunk into R&D results in an $85,000 Cup Car engine that is essentially a collection of almost optimally compatible parts. These engines aren't just "put together" but are the result of a finely honed combination of components having near-perfect compatibility. Notice I did not say "perfect" but near-perfect compatibility. The result is a lot of horsepower but, as any Cup Car engine builder will tell you, it takes enormous quantities of hard work and money to get there. By comparison, we find the engines built by the average hot rodder, even the average knowledgeable hot rodder, are a conglomerate of much less-expensive parts, of which some key parts are far from as good or compatible as the builder thinks.

This is largely due to being ill informed or using a knowledge base derived from advertising say-so. As for the rest of the parts, well, they may be good, but, as often as not, they lack a compatibility common with the rest of the engine components used. The most frequent compatibility/parts selection mistakes in order are cam/valvetrain, compression ratio, cylinder heads, exhaust systems, and induction systems. As you can see, this could result in almost a whole engine being improperly spec'd. For instance, there is no such thing as a "good cam." A cam is only good if it produces performance and that only happens if its event timing is what the engine requires. By the time you get to the end of this book I hope to have reprogrammed the way you think of your engine and the extraction of higher output from it.

Know-How: A Substitute for Money?

First let's define a budget in terms of 2009 dollars. There will be no engine in this book that cannot be built for less than $7,000 with the least expensive starting about $1,000.

Realize this: I cannot cure your budget problem, although knowing how to build engines may certainly help, as your services as an engine builder may become more valuable. What I can do is pass on the technology it takes to build not just 5 or 10 horsepower or more, but 50 to 100 more with whatever budget you may have. A 500-hp-plus "turnkey" engine—by that I mean an engine complete from the air filter through the exhaust system to the tailpipe tips—can be built for under $3,300. The power levels I'm looking at for drag-race applications are up over 600 hp without nitrous, and about 800 hp with, and correspondingly less where an extended service life is called for.

A point that needs to be made and aimed at the over-enthusiastic racer (99 percent will fall into this group) concerns reliability. When building horsepower on a budget it is very important to appreciate the mechanical limitations of the parts being used. For instance, there is little point targeting 600 hp from a normally aspirated 350. This will involve turning 8,000 plus RPM and if you only have stock rods to work with, you will be in trouble. Even though steel cranks and rods are good to 8,000 rpm plus are now options for us, the most successful build will be an engine which makes a lot of horsepower by virtue of high torque output.

Understanding Torque and Horsepower

Let's start by making a simplification. Let's drop the word "horse" because the term "horsepower" just relates to how much power so many horses would make. When James Watt first started building steam engines for mining use back in the early 1800s, he needed to know how many pit ponies one of his engines would replace. Had he lived in India he may have wanted to know how many elephants one could replace. Actually, there's no need for us to tack a prefix onto the front of the word "power." It's still a legitimate term if we just say power. Now let's look at how torque and power are related.

If we exclude the term "horse" then the relationship becomes very simple. Power = Torque x RPM. That's all there is to it. There is nothing more complicated about it. Assuming English units, we measure torque in ft/lb. That is the force times the radius.

Unfortunately, this simplified view of power is out of step in terms of units. Because the force applied to generate torque moves around in a circle we have to bring pi into the equation, but none the less the basic premise that power is essentially T x RPM holds true. If we wish to convert a simple power equation into horsepower, we have to allow for the fact that James Watt decided, by virtue of experimentation, that one horse could do 22,000 ft/lb. of work per minute. Because there may have been stronger horses than the ones he had, Watt upped that figure to 33,000 ft/lb. per minute. Considering this and the fact that the force generating torque moves around in a circle equal to a distance of 2 x pi x R (where R is the relevant radius such as the crank throw, or half the stroke), means we end up with a dividing constant of 5252. So horsepower = T x RPM divided by 5252.

Understanding this can be a great help toward increasing reliability at higher outputs. Read on and you will see why. This simple explanation of the relationship between torque and power is given because it is important that you understand how they are related. Once you have grasped this you will realize achieving your target power is a matter of generating as much torque as possible within the RPM limits imposed by the parts available.

The bottom line is, if we have an engine which makes more ft-lbs of torque at any given RPM, then it will, by virtue of our Torque x RPM rule, make more horsepower at that RPM. Very tight budget constraints usually mean stock connecting rods and these can be a primary limiting factor. Connecting rods mostly break due to RPM-induced tensile loads, but very few break in compression. This means if we're going to make a stout performance engine, it's best done by virtue of high torque over the broadest RPM range possible.

Most automotive materials published in the popular press center around finding horsepower while stressing that torque is really the single most important factor. Unfortunately, and almost without exception, they do not address the issues concerning the generation of torque. In case you doubt this statement just take a good look through the back issues of your favorite magazine and see how often the prime factors of torque generation for more horsepower without the necessity of extra RPM are detailed.

Essentially, I always looked for good torque in my early days of engine building without having anything like the degree of understanding of what it takes to make torque than as I do now. These days I can, with a computer, and experi-ence from 50 years and tens of thousands of dyno pulls, home in on a combination that makes high torque and consequently good horsepower, very quickly. The techniques and know-how to do this are what I'm passing on in these pages.

Builders and Assemblers

Let's consider who will want to read this book. I decided that it would fulfill the needs of two groups. These groups comprise people who are, or want to be, engine builders and people who are, or want to be, engine assemblers. There is a difference and a definition will help establish which group you most likely fall into. First, engine assemblers. These are people who wish to put together parts which are fairly certain to fit together, hitch free, to produce good results. Those who fall within this category really do not want or are in no position to do anything more than the final detailing of parts prior to assembly. About a third of the engines I deal with are engine-assembly jobs. I find the conscientious assembly of an engine returns good job satisfaction and results for the usually limited (compared with "building") effort involved.

For the extra power involved, paying a pro to do the job is not a cost-effective way to go. Another reason for being an engine "assembler" is that most budget-oriented people do not have a lot of work-shop equipment to make many significant changes to the parts used. As a result, like it or not, they become engine assemblers. At the start of any engine assembly project, the idea is to buy cost-effective parts and assemble a functional engine. Engine builders on the other hand, usually are looking for something extra. My pessimistic definition of a professional engine builder is someone who buys a lot of expensive parts that don't fit together. This then means spending hours, even days, making them fit in hopes of achieving the desired goal. Only by applying that extra effort will you gain the racer's edge.

FOREWORD BY ROGER HELGESEN

If you have already read the introduction you may be left with the impression that David Vizard seems a mite more opinionated than most writers of go-fast tech books. Probably so, but let me put that and a few other pertinent facts into perspective. Everything you read in this and any of his many other books is based entirely on results—mostly his. I have, on occasion, accompanied David to some of the lectures he does at various universities to mostly audiences of pro engine builders. A situation that really typifies his approach to performance development springs to mind. A noted engine builder asked his opinion on something. The subject of the question escapes me but the answer given does not. The reply was "I do not have an opinion, I have a dyno!" The inference being that opinions run second best to near irrefutable conclusions drawn and constructed on sound engineering principles from thousands of highly instrumented tests.

David Vizard is a man capable of wearing many hats and successfully does so on almost a daily basis. Because he is so often seen as an accomplished theoretical engineer and innovator, many assume that like so many theoreticians he cannot make anything. Let me set you straight on that one. DV is a very accomplished fabri-cator and machinist. He can competently run just about every machine in a machine shop from boring machines, mills, and lathes to crank balancers and of course dynos and flow benches. If something he wants is not available, or is beyond his finances, he makes it. In his early days, he filed a cam for a four-cylinder race engine out of bar stock. He also made the front spindles/steering arm setup for one race car by starting with two solid-steel billets, 9 x 9 x 6 inches, and hack sawed and filed the end product into shape! Sounds like a lot of effort—this begs the question as to whether such efforts to improve engineering skills, to sort fact from fiction and to apply what is learned are worth the results.

Here are some results; judge for yourself. Let me set the scene: It's 1971 in DV's home country—England. The venue is Prescott, one of England's oldest race tracks. This 1100-yard course winds its way through scenic English countryside and climbs some 300 feet from start to finish. A very short, asphalt paved version of Pikes Peak hill climb, if you will. At the particular meet in question DV beat all the sports cars (Cobras included), sports racing cars (GT 40s included), and Formula 2 cars. He also beat the 180- to 200-inch four-valve-per-cylinder, fuel-injected, 11,000-rpm, megabuck Formula 1 cars with a 78-inch, two-barrel carbureted Austin Mini Cooper that cost less than the rear tires of the quickest F1 car he beat!

This sounds to me like someone who not only knew what he was doing with engines but also with brakes and suspension systems and on top of that knew how to drive. This was no isolated incident. In one season, from just a few engines built, DV scored a combined 169 track records, pole positions, race, and championship wins. Five of these engines made national champions, two seconds and a third. Not bad for someone who has considerably more than a fulltime job writing.

As a driver DV has held class records on every major racetrack in England. An incident at an event the press introduced new car models ahead of time is worth relating. GM brought along an IROC Camaro and a mildly hopped-up production Camaro for drivers to try their hands on a huge expanse of asphalt at GM proving ground. Although he had no previous experience in a race car of such size and weight, DV set fastest time in the IROC car and third fastest time overall in the street Camaro. In other words, he beat IROC times of all but one driver in a street car!

An incident at the main supporting race of the 1974 British GP is worth telling. After a combined practice and engine break-in session had placed him 28th in a 32-car field, DV came into the first bend between two other cars, each having three times the engine displacement!

So you want cutting edge technology? David Vizard supplies it; I can back that up in no uncertain terms. Try this for size. In 1991 DV and his friend David Anton built a 2000cc Ford Pinto engine for a Pro 4 circle track car. The engine spec came right out of DV's 2000 Pinto book published in 1981, so in essence the engine was a 10-year-old DV spec. It was put into a race car built from the front and back half of two crashed cars. The driver, though experienced, had never won a race.

On the track, this combination proceeded to pull away from the track record holder and current champion, a professional race car builder, by half a car length per straight. Can you imagine how frustrating it must have been for the champ's nationally acclaimed engine builder? Ten years had gone by and he had not equaled DV's part-time development efforts!

On a bigger scale, David achieved something along the same lines while consulting for Chrysler UK. For Chrysler's entry in the British Touring Car Championship, a series fought almost exclusively by manufacturers from all over the world, DV modified the heads, did a substantial redesign on the Weber carbs and advised on cam and exhaust specs. Even with a dated pushrod engine against a field of OHC competition, Chrysler cleaned up two consecutive years with 26 wins in 26 races!

While I have read many automotive writers who are gifted with words, it is clear they have not spent the thousands of hours practicing the art of finding power that so plainly makes the difference. Most writers get their information from industry professionals who may have a narrow focus on the way many

components interact. Your race-winning combination requires an integrated combination of components and determining those is very much a DV contribution to motor sport technology.

One of the unique traits of DV's writings is that he reveals the latest technology and even points the way to future developments. As an example, David's daughter raced a mini she and her boyfriend built using his 500-page book *Tuning British Leyland's 'A' Series Engines.* For the two-year period it was raced, this car was unassailable. It thoroughly thrashed every Mini it raced against whether professionally built or otherwise.

This 1275 Mini GT achieved more than an 80 percent win record and was never beaten by anything of less than 3100ccs! His only connection with the car was writing the book. The closest he ever got to the car was 5,000 miles. This is only one example of his far-ranging engineering expertise. Since this is a Chevy book, let me relate something more to the point and throw in a caution concerning HP numbers and comparisons.

David's close friend Terry accumulated a heap of parts from frequent swap meet visits. From these parts he had built a motor which was installed in a '69 Camaro stripped of anything not needed for drag racing. With a freshened block, new bearings and rings the car went a very creditable 10.2 seconds at 133 mph for the quarter mile, which calculates out to about 460 hp. While the parts were somewhat limiting, David's machining, assembly and dyno setup brought this up to 534 hp. With a race weight near 3,000 pounds, Terry's Camaro ran a 9.82 at 139.3 mph—at Carlsbad, a track not noted for traction. Not bad considering the complete shooting match cost less than $6,000 (in 1991 dollars) to build.

My point is that this car blew the doors off cars which reportedly had much more power. An example being a big-block car that at a well-known southern

California race shop had dyno'd at 650 hp and could only manage 10.4 seconds and 135 mph on the same day. Time after time, DV's engines beat others with supposedly more horsepower. The point is that engine builders sell engines by virtue of the amount of power they can claim of them. Except on rare occasions, DV does not sell engines to the public but he does sell facts.

Since David Vizard's dyno results are seen and scrutinized by so many worldwide, calibration of both dyno and sensors is a primary consideration. This is not necessarily the case with many busy commercial engine builders. The point here is that you need to be careful of making horsepower comparisons. If you build by David Vizard's books, you will, in most cases find, that you will drive right by a lot of cars with supposedly more horsepower.

I could easily fill ten pages with information about DV that you're unlikely to read elsewhere. Why? In the unlikely event he was looking in DV's direction for a story, why would one automotive writer want to write about another? That's a little like promoting your business opposition.

It might appear that if it is done competently enough, it could adversely impact that writer's own business. But there have been notable exceptions—in his home country, where he is known as Vizard the Wizard (in the USA he is known as Mr. Horsepower), a number of writers, impressed with DV's accomplishments, have made him the subject of some very complimentary editorial instead of the producer of editorial.

With almost 4,000 articles and 30 books, DV's material has much to offer you. When your intent is to build a successful competitive machine, David Vizard's books represent, in my opinion, the best choice you can make. My advice is use this book to its fullest extent and go win some races.

ENGINE BUILDING PRACTICES

The Importance of Good Parts and Engine Building Practice

A good place to start is the emphasis on good engine-building practices. Any time you're working with limited finances, you are forced to compromise. This makes it even more important not to compromise your engine build in other areas. The most important asset that you can control is good workmanship. When circumstances force the building of an under-financed engine, you should put in your absolute best effort to build the best engine within your budget.

The number-one requirement when assembling any engine is a clean work place, along with clean tools and clean parts. The second rule is to never leave your engine building to the last minute. The more compromises you're likely to make, the more chances there are for mistakes—that's guaranteed.

A mistake made during assembly causing a subsequent teardown because of broken parts means your budget has taken even more of a beating than you expected. That, I'm sure, is not your aim. If you're new to engine building, everything will take at least twice as long as you expect. If you can find the patience

Having the components for the build is only part of the formula for a successful engine. The other essential element is having a clean and tidy work area like you see here.

and skill needed to put an engine together, it will pay dividends. I'm avoiding the often-used term "carefully" because it describes nothing of any consequence. Let's remove the word from our performance dictionary and replace it with the words "conscientiously," "thoughtfully," and "intelligently." If these are applied, the chances of achieving good results are multiplied.

For instance, when building a motor to run in stock classes, I have—in an effort to get the best parts—built fixtures that measure rocker ratios. With this I went along to the dealership and sorted literally thousands of rockers until I had a set that gave me the highest ratio. That is just one instance. On another occasion I managed to wrangle my way into the factory that made the vehicle I was racing. I parked myself on the engine-assembly line for two days measuring critical block dimensions with micrometers and bore gauges until I found a suitable block, which I then left with. Was it worth it? You bet.

This and many other legal moves made me unassailable on the track. I fully understand you may not have the time or resources to do this. However, nothing should stand in the way of doing your best. As for the question of what good workmanship is and how the engine should best be put together, well, there are dozens of facets that need to be covered. Where they are directly applicable to what we're doing is dealt with at the appropriate time throughout the following chapters.

Next, consider how much money you have to spend. When it comes to performance, there's no such thing as an over-financed engine. I remember asking a team manager friend of mine why he felt that competing teams always managed to just beat his team into second place. Straight-faced, he turned to me and said, "There's no substitute for cubic money." Ordinarily, one might expect there was a degree of sour grapes, but this was far from the case. Knowing this gentleman and knowing the team he was up against I realized his cryptic statement was valid. There is no substitute for cubic money. But throughout this book I'm going to reduce the effect of limited funds by advising how better to spend the money you do have.

Hardly a day goes by when I'm not confronted with an engine being put together on a limited budget. Usually, money has been spent needlessly on machining operations and parts that do nothing to aid performance or reliability. Also, when you start building your engine it's a good plan to have a realistic idea of what the parts will cost compared to what you can afford. In the real world they always are more than anticipated, so, to avoid disappointment, acquire current catalogs or go online and check prices from companies that stock a wide range of parts. Companies I recommend are Summit, Jegs, Nichols, and Scoggin Dickey. These are nationally known companies and are always on the cutting edge in terms of pricing. This doesn't mean their prices are the lowest you'll find. What it does mean is that they offer a range of products at cost-effective prices and a money back guarantee.

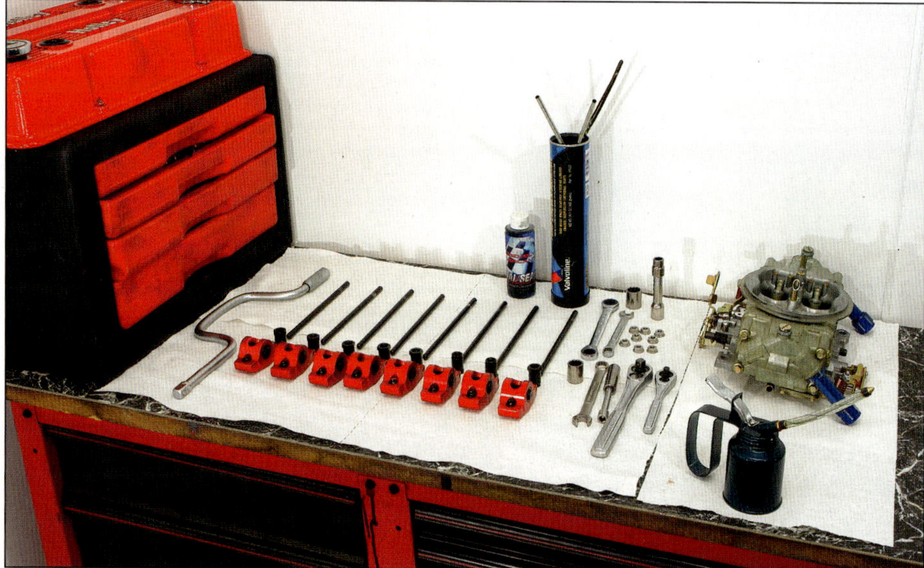

A clean shop encourages clean parts. There is little point in cleaning the parts if you don't have a clean place for them afterward.

By dealing with a company specializing in certain parts, you may be able to get prices marginally lower. Admittedly, it's unlikely the reductions will be drastic, but if you don't have much money, any budget cutting is a help. Of course, there will come a time when the price of a non-brand-name part looks too good to be true. When that happens, you can bet that it usually is. An economic measure is to make sure you buy from a reputable company that will stand behind its parts or workmanship. Nothing will be more expensive than having an incompetent machinist do your work or a supplier who sells you parts claimed to have certain qualities that they don't and instead are second-rate replicas. It does happen, take my word for it.

Up to this point I've worked on the premise that you're going to buy new parts. There's also the option of buying used parts. Many used parts do not suffer from degradation simply because they've been used. For instance, you can often pick up a used intake manifold at half the price of a new one. As long as it hasn't suffered any severe corrosion in the water passages, it will function as well as new. The only real factor you have to worry about is whether the manifold is any good.

Paper towels can leave lint, but this is infinitely better than leaving grit. Use as many clean paper towels as the job requires.

So much for parts that don't have any moving components within them. They obviously aren't about to suffer from significant wear degradation. What about parts such as connecting rods, pistons, camshafts, valvetrain components, and so on? These obviously experience wear, but if you buy wisely you can get good parts at a good price. The question is, what's good? Some parts you need to steer clear of unless you know what you're doing. Be sure the parts are sound before buying and installing them.

Pistons, although they're aluminum components, can present a different picture. Close inspection of a set of pistons can reveal much. If the skirts, ring grooves, and pin bores pass a thorough magnifying-glass inspection and no cracks or unduly worn surfaces are revealed, you're in business with a functional set of pistons. Functional means being in one piece. Whether they're the best you could have bought for making power may be another thing.

Swap meets, though, are not the only source of used parts; often you'll find parts in local ad sheets, and a source often untapped is your local speed shop. These days the Internet has also become a useful source of used parts. Although I have not gone that route myself, I do have friends who have built some great engines using parts they have acquired from Racingjunk.com, eBay, and a few other reputable sites. However, you need to ensure the parts and source is reputable. The last source I'm going to mention for buying parts is the wrecking yard. I've left this until last because it can be a gold mine. If you develop a working relationship with a wrecking yard, it can be an extremely good source for parts.

For example, a dealership may be closing or a company with a one-make fleet of vehicles is changing the model or brand it uses, so its spares are useless. I've managed to get several new steel Chevy cranks this way at about a quarter of the dealership cost, and once even bought almost-new Corvette aluminum heads for $125 for the pair—a bargain by any standards. To get this kind of price, contacts are everything.

Inspection

Whenever a motor build is budget constrained, the emphasis shifts from the purchase of expensive, quality parts to inspection and quality preparation of parts built down to a price. This means the responsibility for ending up with the best parts possibly shifts from the component manufacturer to you, the user. For example, if you're in the market for a set of race rods and money is no object,

Some operations have to be farmed out to a competent machine shop. A shop that delivers fast, accurate machining at a fair price is one of the most valuable relationships you can cultivate.

Optimum ring seal and proper bearing clearances will count for naught if cleanliness during assembly is compromised. Here ARP head studs are being moly greased prior to dropping the head on.

Choosing the right cam is of paramount importance. Chapter 7 precisely explains what is needed.

then you'd expect your newly acquired Oliver, Scat, Crower, or Carrillo rods to be close to perfect right out of the box. You will have paid enough for them and rightfully can expect them to be as close to perfect as modern engineering allows.

The same can't be said for the stock rod. This is produced down to a price and is designed to withstand loads commensurate with a stock motor. For many of us, the application for which we intend to use it will, to some degree, overload it. If we are to make an average rod survive as long as possible, we'll have to consider whatever moves may be open to us to improve the rod's survivability.

The first move is always inspection. Only after you're sure that no obvious

Your engine may never see a dyno, but strict prep and assembly practices will ensure positive results along with good reliability if you adhere to the information given in the following chapters.

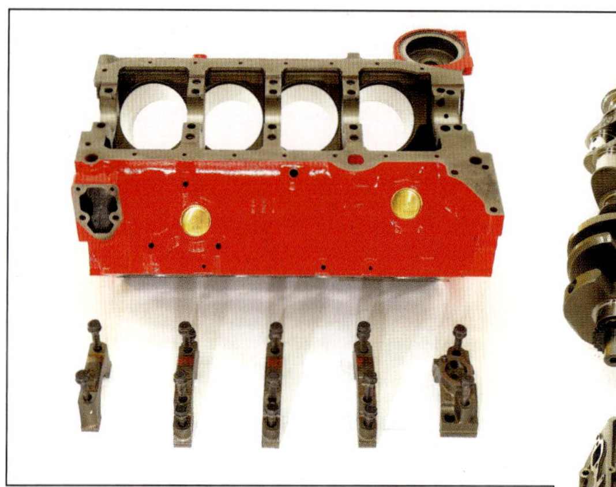

All these parts must fit and function correctly to be successful. Be sure to inspect every one of them.

component flaw exists should you accept the part for your engine. Sure, finding one bad component in a set is inconvenient, especially when you had planned an assembly session. Now you have to wait until you can locate another part later in the week. But the inconvenience and expense is far less than the inconvenience and expense of having to pull a broken motor from your car.

Make inspection a religion to the fullest extent allowed by your tool kit. Obviously, such things as micrometers cost money and may not appear on the list of tools you can afford now. This does not preclude you from using your eyes and simple checking techniques to establish, within reason, that parts are acceptable. No matter how good your machine shop may be, sooner or later a mistake will be made.

The problem becomes personal when the mistake happens on your block or other part. Parts inspection must be part of your build program long before the components concerned go for machining. Only after you've established that the components are fit for the application should you begin parts preparation.

Preparation

The term preparation is, in the context being used here, somewhat broad-based.

This usually involves metal removal by one means or another. In its simplest form, the metal removal most often will be by means of a fine file or emery cloth. At the other end of the scale, the preparation operation involved may be so extreme that it requires relatively expensive machinery. An example is the lightening of a stock crank. This involves setting up in a milling machine and drilling the big end journals hollow to allow a counterweight reduction for lower windage losses.

Preparation work of such a magnitude more accurately might be called a modification, but we could say that parts preparation is really only a question of minor modifications. Whatever we choose to call it, it doesn't make a lot of difference. The result is that preparation makes a difference in both reliability and performance.

At the end of the day, probably the best definition of preparation is: It's the amount of effort you're prepared to put into the parts involved. Where material removal is involved, the most basic parts preparation tools are a simple set of inexpensive needle files and some fine emery cloth.

Next on the list is a 6-inch dial caliper with enough precision to detect possible machining errors and critical assembly dimensions. After this, a die grinder is the next most important parts-preparation tool. This allows a great deal of horsepower-generating parts preparation to be done. The most productive is the preparation of the cylinder heads, which easily can account for 30 but could

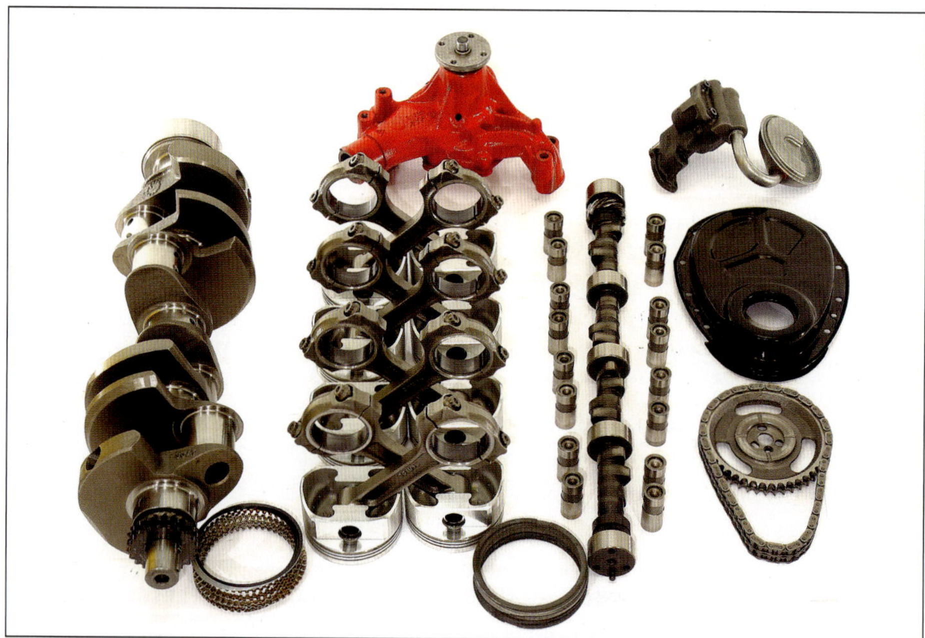

Here are the principle components that go into the block. Just how much power potential the engine has ultimately depends on the choices made of crank, rods, and pistons.

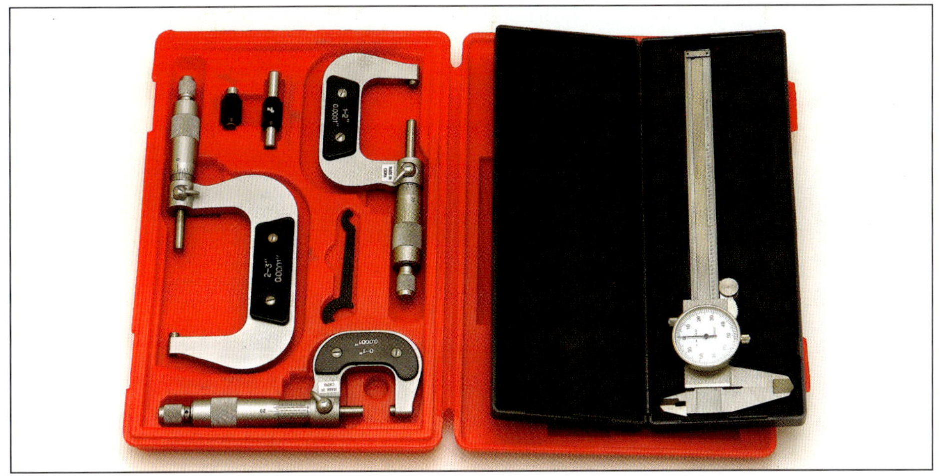

Many parts depend on close fits for trouble-free operation. A set of inexpensive micrometers and calipers from ENCO are a great help here. The most important for a small-block Chevy are the 0- to 1-inch and the 2- to 3-inch items. Don't worry about buying the less expensive offshore-produced brands. If you are just building engines they will last a lifetime if treated with reasonable care. The same goes for the dial calipers.

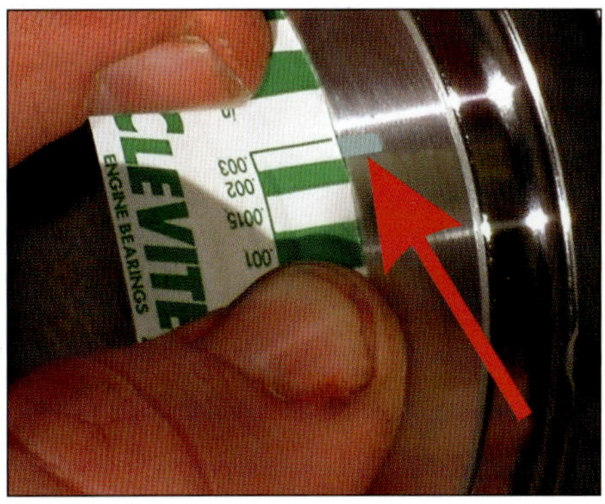

If you have wondered how you will measure crank bearing clearances without an expensive dial bore gauge, then worry no more as there is always Plastigage. This is about as cheap as it gets for clearance measurement.

be as much as 80 to 90 hp. Power achieved by preparation takes time but costs little or nothing in the way of cash. It can be applied to a variety of parts such as blocks, cranks, rods, heads, manifolds, carbs, distributors, and the list goes on.

The break-even or payback on a die grinder is relatively short. Once you've acquired a die grinder you will have reached a plateau in terms of tools. Accumulating regular hand tools such as wrenches, a drill gun, etc., is your principle priority. However, there comes a time when having a small lathe and a mill really starts to pay off. For most of you that will be down the road quite a ways, but the payback on a cheap, Bridgeport-style mill is at about the 15th-engine mark.

Although I've put a great deal of emphasis on it, parts prepping isn't essential if maximizing horsepower isn't a prime requirement. In some instances, dedicated component preparation may be frosting on the cake. In the introduction I talked of builders and assemblers. An assembler, even a good one, needs only perform the inspection aspect and correct whatever errors are discovered. If you're assembling an engine that's expected to put out no more than that of a healthy stock motor, detailing the parts may be of no real consequence. However, to make that transition from a good assembler to an engine builder—and maybe on to a successful professional race engine builder—essentially relies on your ability to effectively inspect, prepare, and detail components. Remember that before you dismiss anything as inconsequential.

Now that we have the main aspect of parts preparation over with, let's consider the helpful little tricks that can considerably extend the life of components. These come under the heading of good assembly practices. Little things need attention, like smearing oil on the lip of the crank seal so it doesn't see a dry start and so on. If you have never put together a small-block Chevy before, then take note of the assembly tips given throughout this book.

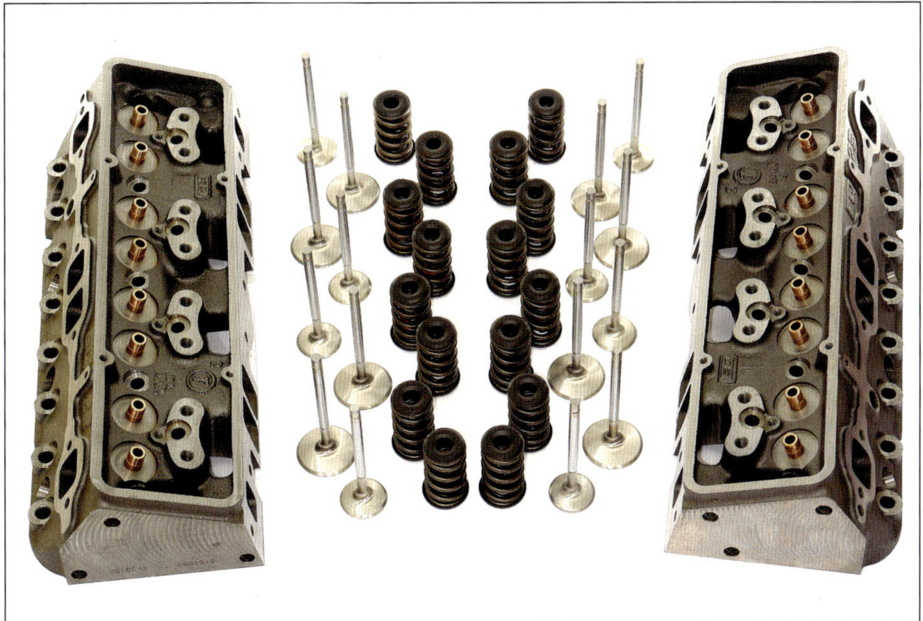

Detailing head components always pays off. See Chapter 6 for more information.

THE PRODUCTION OF POWER

If you were to read this book without realizing that in almost every instance the goal I'm trying to achieve is more horsepower, you'd think—and rightly so—that I'm obsessed with airflow. Throughout the chapter on ram charging, air filters, carburetors, induction system, cylinder heads, and exhaust, the airflow potential of the various parts are discussed at length. I assure you this, as you will see when we delve into the details of engine performance, it is no obsession. Airflow is the prime ingredient for maximum horsepower.

Here is a prime example of a low-budget car. That, however, did not stop it from posting competitive performances.

Flow Bench Relevance

Let's start off with the basic question: Why do engines have carburetors? The simple answer is that they're there to mix fuel and air. Why is it necessary to mix fuel and air? Again, the simple answer is that fuel will not burn on its own; it needs oxygen. The oxygen is acquired from the air drawn into the engine. For a given amount of induced air the engine can effectively burn a given amount of fuel. It's the action of burning the fuel that causes the gases to heat up and expand. This increases the pressure in the cylinder, thus pushing the piston down the bore on the power stroke. The greater the amount of air drawn into the cylinder, the greater the amount of available oxygen there is to burn with fuel. The more fuel that can be burned, the greater the amount of heat generated and, therefore, the higher the pressures generated in the cylinders. Greater pressures mean higher horsepower by virtue of higher torque.

The bottom line is that the more air the engine can inhale during each induction cycle, the better off we are for extracting power. Breathing efficiency boils down to using components that allow air to flow as freely as possible into the engine, hence the apparent obsession with airflow capability. Any component that doesn't flow air will effectively rob the engine of some of its potential as a power producer.

Heat Management

An internal combustion engine is so named because it burns fuel and air internally. It does this within the working cylinders. An example of an external combustion engine might help you to see why we call a gasoline engine an internal-combustion engine. The most obvious example of an external combustion engine is a steam engine. Combustion of the fuel takes place outside the cylinder

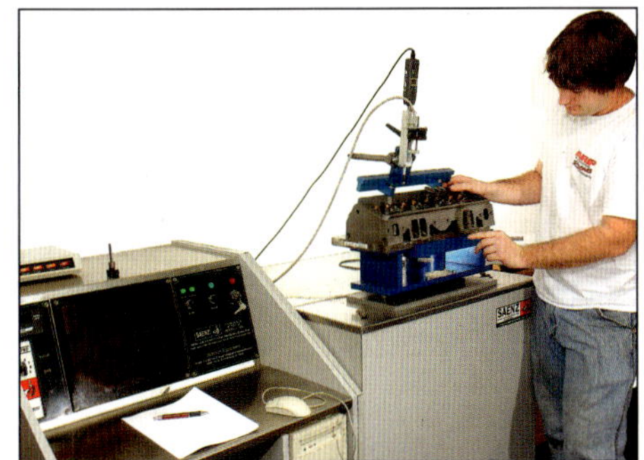

Here's the flow bench currently in my shop as of 2009. It has enough capacity to flow all 4 barrels of even the biggest of 4150 series Holley carbs. It will also pull up to 120 inches of water. The device on the heads is an Audie Technology auto valve opener. This allows us to cycle through a test really quickly.

in the furnace that heats the water in the boiler. Since there are few steam-driven small-block Chevy's around, I'll skip anything further on that subject.

Going back to the internal combustion engine, we find that heat produced by burning fuel causes the gases in the cylinder to expand. Incidentally, the gases don't explode. The process is far too slow for an explosion; it is a burn. It is the heat-induced expansion of the gases creating pressure on the piston and pushing it down the cylinder that develops power.

The fuel has to be mixed in well-defined proportions in order to burn efficiently. Although there are various valid reasons for working above and below the chemically correct ratio of fuel and air mixture, we can say that burning the fuel at the chemically correct mixture will provide good results.

Fig 2-1 shows the approximate proportional volumes of fuel and air consumed in one minute by an engine developing around 400 hp. Drawn to the same scale are the eight intake valves the mixture must pass through in the time they're open. Assuming a 300-degree race cam, the intake valves are only open for 25 seconds of that one minute.

The valves look small compared to the volume of air that must pass through them. The implication is that having them flow efficiently is of prime importance.

We know that the heat in the cylinder causes the gases to expand and this in turn pushes the piston down the bore. Unfortunately, we can't use the whole potential of the heat energy produced during the burning cycle. This is due to the nature of the engine's design and the fact that the cylinders lose heat in many areas. The engine has to vent the cylinder to the atmosphere by way of the exhaust valve long before the cylinder pressure drops to atmospheric pressure, which means a big loss of energy. If you refer to the Fig 2-2 on page 16, you'll see where most of the heat losses are occurring.

It's essential at this point to realize that heat energy is directly related to horsepower. Assuming a 100-percent conversion efficiency, it takes 778 British Thermal Units (BTUs) of heat energy to develop 1 hp. Alternatively, it takes 1 hp of mechanical energy to produce 778 BTUs of heat energy. It's more practical, though, to convert mechanical energy into heat energy than the other way around.

For this edition of the book, a lot of the dyno testing was done on this DTS dyno. The dyno cell was a high-tech, environmentally controlled one so as to minimize correction factors.

For instance, when we dump a certain amount of a vehicle's kinetic energy into its brakes, they turn all of the absorbed kinetic energy into heat at a 100-percent efficient conversion rate. At the end of the day we find that, out of the potential energy from the fuel burned, the amount of energy actually extracted in terms of power at the flywheel is limited.

In fact, for every 100-hp worth of fuel burned in the cylinder, a good engine will derive only about 25 hp at the

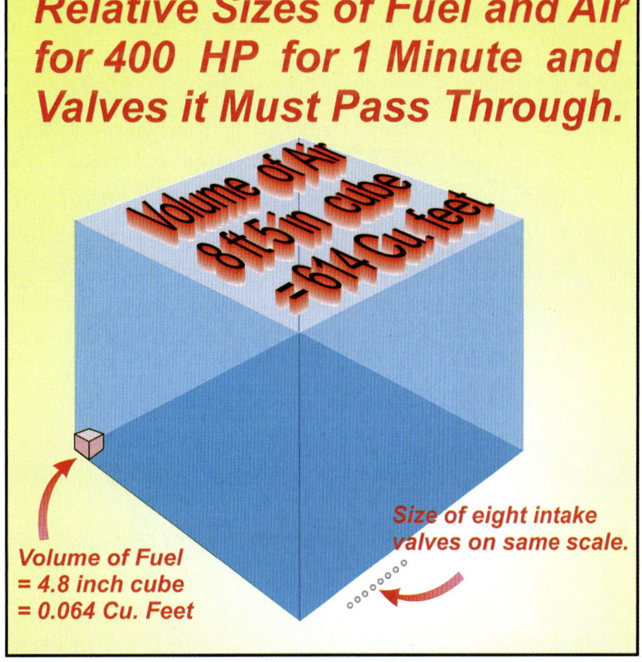

Fig 2-1. What you see here is the amount of air and fuel for 400 hp for one minute drawn on the same scale. Also, the size of the intake valves are shown on the same scale. What is immediately apparent here is how small the intake valves are in relation to the amount of air they have to pass into the engine. This should amply demonstrate the need for heads that flow air well.

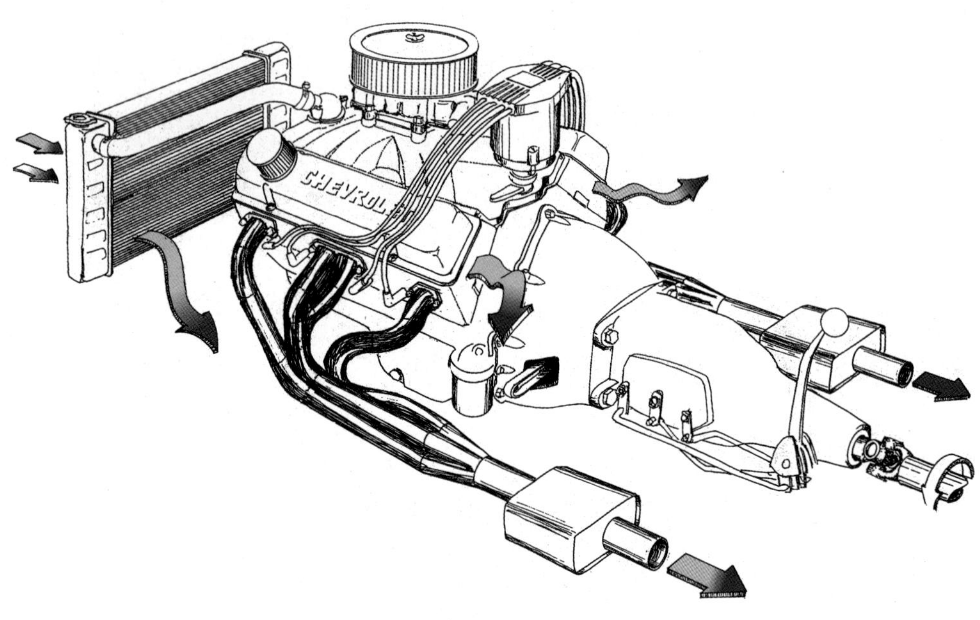

Fig 2-2. From this illustration you can see that of the fuel burned only about 25 percent of it is actually converted to mechanical power. Some 75 percent is spent heating up the atmosphere. With due diligence we can improve the engine's fuel efficiency.

flywheel. This rate of energy conversion of the fuel's potential energy into flywheel horsepower is known as the engine's thermal efficiency. A figure of around 25 percent is typical for a good engine. A normal road engine is often around 18-percent thermally efficient.

Examining the cylinder pressures that occur within the engine, we find that the power produced from the cylinder pressure is more than that seen at the flywheel. The difference between these two numbers is a measure of the engine's mechanical efficiency—that is, its loss of power from friction and pumping losses. Unfortunately, horsepower lost to friction is turned directly back into the heat that's carried away by the cooling system.

When an engine is modified, attempts are made to minimize all of these losses on one hand, and to improve the rate at which the engine consumes air on the other. As has already been demonstrated, the more air and fuel mixture that can be passed through the engine at a given time, the more the power output will be. This of course only holds true so long as none of the other inefficiencies are unduly increased. If this is achieved, then the engine will show more power, and we determine whether power has been increased on the dynamometer. A dynamometer, or dyno as it is more commonly called, is a device for measuring horsepower. There's a lot of confusion about horsepower and rating numbers, but we'll get to that later. At this point let's look at what it takes to make horsepower.

Although airflow has figured strongly in our discussion so far, it would be wrong to think that airflow is the sole key to horsepower. It just happens to be one of the most important ones. Like any complex device, if we fail to produce results in one area then the overall results will be less than hoped for. There are factors other than airflow to consider.

To more easily understand what and how these factors affect the overall scheme of things when discussing various topics, refer to the power production flow chart on page 17.

With the aid of that chart, we'll go through major factors that must be considered in any plan to develop high output, not just from a small-block Chevy, but from any engine.

At the top of the chart we start with atmosphere, which is the prime ingredient that must be moved through the engine in as great a quantity and as efficiently as possible. The atmosphere supplies the oxygen that allows the burning of fuel. This in turn generates the heat that expands the air that—since it's contained in a closed cylinder—rises in pressure and pushes the piston down the bore.

Obviously, the greater weight of charge the engine inhales, the greater it's potential for power. This brings us to step two, which involves maximizing the air density passing into the engine. In our example, it's difficult to cool the air below that of the prevailing (ambient) temperature of the surrounding atmosphere. It's important to understand that it's not cubic feet per minute (CFM) that generates high output but pounds per minute. Hot air expands and weighs less, and consequently contains less oxygen, than cool air. As a result, it's worth the effort to see that the induced air isn't heated more than necessary, thereby maximizing its density.

The next step involves minimizing intake flow restrictions. Here's where flow bench work on the induction system, intake port, and (to a certain extent) combustion chamber wields considerable influence. If ever a factor needs emphasizing, it's the total induction system's flow capability. Getting a charge to effectively fill the cylinder is the single most difficult function to achieve in the design and development of an engine. Never lose sight of this while speccing and building your engine.

In terms of induction effectiveness, we have to maximize pressure wave tuning.

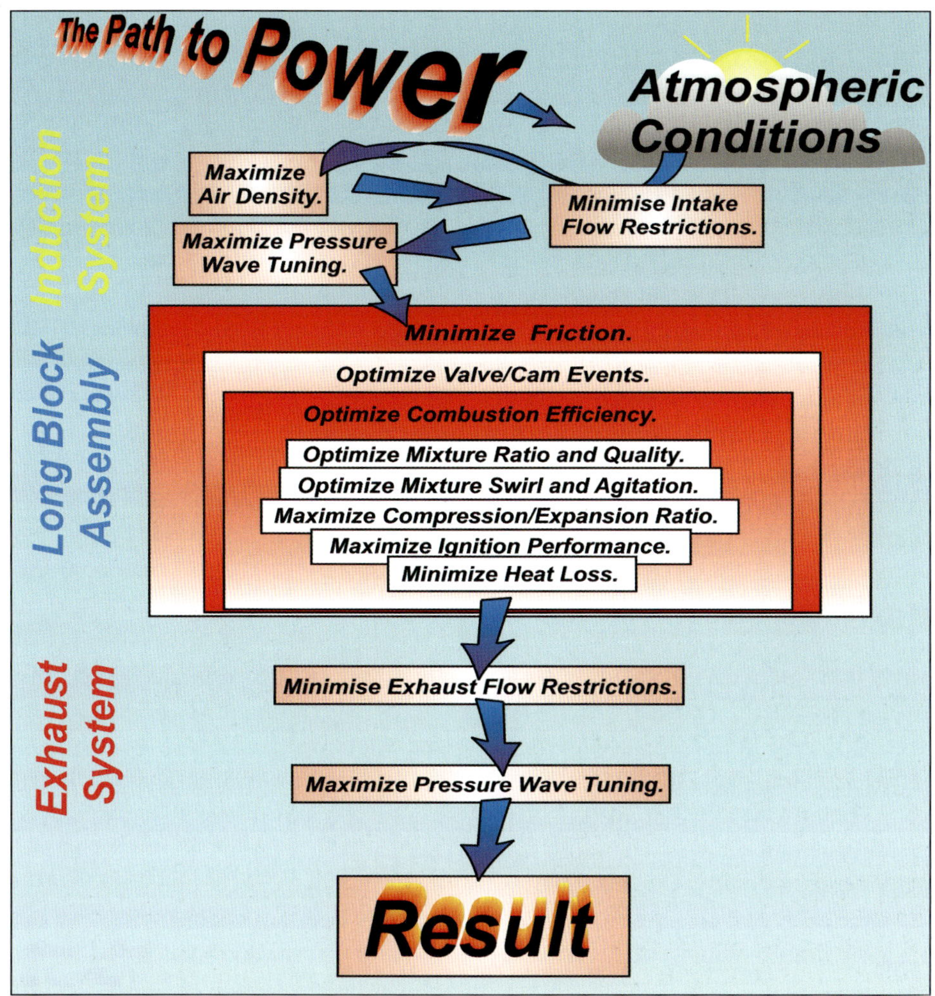

Although the effects of this cannot be quantified on a conventional flow bench, measurement of airflow on running engines does demonstrate its importance to power output. If taken to the limits of current technology, then, with a well-developed induction and exhaust system, volumetric efficiency (breathing efficiency) figures well over 100 percent can be achieved. These results are from the combined effect of both intake and exhaust pressure wave tuning. At this moment we're considering only the intake, which in practice operates over a relatively narrow power band. But when combined with the stronger and more effective exhaust tuning, the entire system becomes far more effective.

Now we come to the engine, or to use a more common term, the "long block." Our overall concern with the long-block assembly is friction. Everything within the long-block assembly needs our attention in terms of friction reduction.

By paying attention to friction reduction, not only do we allow the engine to make more horsepower, but it also lasts longer. Just for the record, the effects of friction become increasingly detrimental as RPM increases. Just 10 ft-lbs of additional friction within the engine (an easy amount to incur) will cost 4.8 hp at 2,500 rpm, 9.6 at 5,000, and 14.4 at 7,500.

Contained within the long block and the friction box in the chart above is the valve/cam event box. This refers to when and how high the valves open. It's an important and critical factor that must address both the intake and exhaust requirements simultaneously. To be successful at intake and exhaust, both have to happen at the appropriate time—in relation to the crankshaft rotation. This makes cam events more of an overall factor rather than something pertinent to the intake event alone.

It's worth mentioning that the subject of optimizing cam events is one of the least understood areas of high-performance engine building. Fortunately I have some pertinent information in that area because it's a specialty of mine. If you take the time to absorb—even in its simplified form—what's covered in the cam selection chapter, you'll be better than one step ahead of the opposition. Guaranteed!

Now we're getting down to the real core of the process of the heat engine: combustion efficiency. In the combustion efficiency box there are five factors that we need to deal with. Failure to optimize any one of these factors means reduced torque, which in turn means less power. To avoid failure, be sure to heed (to the letter) what I have to say on the selection of production heads. For a small-block Chevy, heads are a major issue, but in other areas the output may only suffer minimally if a component is a little off optimal.

The first combustion efficiency factor, the mixture ratio, has to be closely controlled within narrow limits and the mixture quality has to be what the engine wants. Here the mixture quality refers to fuel/air mix with the fuel sufficiently atomized for the engine's combustion requirements, but not over-atomized so as to cut volumetric efficiency. Establishing the best mixture quality for power or economy is a major issue for engines requiring a wide operational band, as all good street motors do. Because of its importance, we will deal with it in detail in the carburetion, induction and cylinder head chapters.

The second factor is mixture swirl and motion. Fortunately, the typical production small-block Chevy intake port has reasonable swirl characteristics.

This is a good start, but the type of combustion chamber a production head may have can make or break any assets the intake port may have. Combustion effectiveness will also suffer if the block

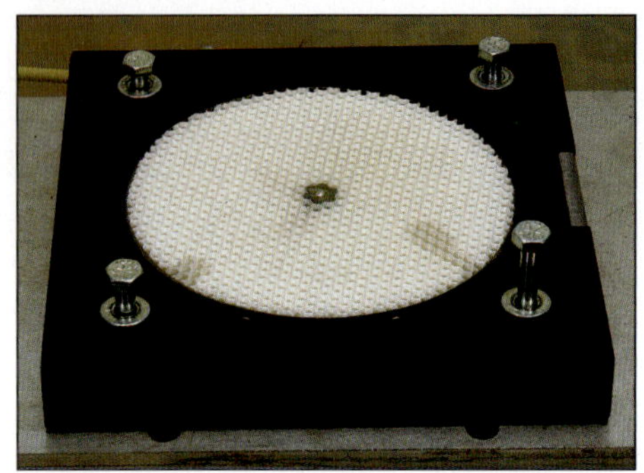

A swirl meter can make important contributions toward developing heads that produce a wide power curve. I have used three types over the years: a paddle wheel, a torsional torque, and the honeycomb-style one like the Audie Technology unit shown here.

assembly is built incorrectly, because the piston-to-cylinder-head-induced squish action hasn't been optimized. Failure here causes a drop in torque, horsepower, and fuel efficiency. On top of this, your engine will be more prone to detonation.

Third are the compression and expansion ratios. I'm sure most of you are familiar with compression ratio, but probably few are familiar with the term "expansion ratio." Expansion ratio is simply the reverse side of compression ratio. When we raise the compression ratio, we're more interested in what it does to the expansion ratio, because it has a significant effect on our optimal valve event timing and the ratio of the cylinder heads' intake and exhaust valve sizes. We'll go into sufficient detail in the relevant chapter to arm you with the working knowledge needed.

The fourth factor in our chain of events is maximizing ignition performance. Here we have to ensure the engine has more than just an adequate spark.

Once a more-than-adequate spark is established, any extra spark output will not return a further cost-effective output. High-performance ignition systems will boost output, but by only a relatively small amount. However, anything short of adequate will cost substantial power.

The fifth and last factor in the combustion efficiency box is the minimization of heat loss. This is an area that's been given too little attention in the past, mostly because the materials to improve it were not really available. They are now, so we'll look at minimizing the combustion heat loss from the chamber so that more heat is available to expand the air and thus deliver more horsepower.

We have now passed through the friction and combustion efficiency boxes and into the exhaust system. Our first goal, like that of the intake, is to minimize exhaust flow restriction while maintaining cross-sectional areas and lengths appropriate for effective pressure-wave scavenging of the combustion chamber. This may seem like an easy aspect to deal with but, judging by the number of incorrectly spec'd exhaust systems I've seen, this must not be the case. With currently available parts, it's entirely practical to build a "no-loss" muffled exhaust system that meets even the most stringent street standards. On top of that, it also need not cost you a fortune. It's little more than buying the right parts and then building the system correctly.

The exhaust system actually starts at the exhaust valve, not the exhaust manifold. Restrictions can be reduced in a number of ways; the obvious one is to flow the system, but another is to make sure that the valve is open for as long as possible to get the job done without compromising the effectiveness of the power stroke. This is an important point to appreciate because it's greatly influenced by the compression and expansion ratios. Understanding the implications can make the difference between 1.5- and 2.5-hp per cubic inch. Granted we won't make the 2.5 end of the scale with the hardware allowed by our budget, but the statement still holds true in terms of what it takes to achieve such an output from a naturally aspirated engine.

Last we come to maximizing pressure-wave tuning of the exhaust. This is an

Here is one of our Performance Distributors test units. Instead of a vacuum can it has a micrometer advance retard knob (middle-left in photo).

important factor. It's not commonly realized that exhaust pressure wave tuning has a far greater effect on engine output than the intake. A normal intake system with one port per cylinder is difficult to tune over a bandwidth of more than about 400-rpm wide. However, if the exhaust system is done correctly, it can be made to effectively scavenge the combustion chamber of spent charge and pull in a greater charge over a power band as wide as 4,000 rpm and sometimes more. In the following pages, we're going to deal with all of these things for a better understanding of how they interact and how they fit

Here is where Performance Distributors sets up the advance curves for their customers' engines. This rig is also used to check the RPM capability of the HEI-based ignition systems.

Here, performance engine builder and tech writer Bob McDonald runs the numbers on one of our project engines. This 355 turned a healthy 506 hp and 461 ft-lbs and it was really affordable.

together to make what an experienced engine builder calls "the right engine combination" to make horsepower.

Dynamometer Results

Throughout the pages of this book you'll see tests of an engine's output. These figures were developed mostly on my dynamometer. This is a device for measuring power. There are generally two types of dynos in common use: an engine dyno, and a chassis dyno. The engine dyno tests the engine while it's out of the vehicle. The output at the engine's flywheel is transmitted directly to an absorption unit.

A chassis dyno measures the horsepower at the driving wheels of the vehicle. This is done by simply driving the car onto a set of rollers that are coupled to an absorption unit. Whether it's an engine dyno or a chassis dyno, a dynamometer is merely a device for applying a braking load to the engine. Hence, the term: brake horsepower (bhp).

Rating Standards

When looking through car catalog information sheets, you no doubt checked the quoted horsepower figures. Survey enough specifications and you'll see "DIN" figures. Others are marked as SAE "gross" or "net," and yet another as "standard" horsepower figures. Through the years, manufacturers have obtained horsepower figures from blueprinted engines or with open exhaust, unrestricted air intake, and no accessories hooked up. The results are unrealistic horsepower and torque figures.

As for the power figures quoted in this book, they're known as "standard" corrected numbers. These numbers are a little more flattering (around 4 percent) than SAE figures, but not without at least some justification. They end up producing a higher corrected number because they correct to a higher air density and a cooler intake charge temperature than the SAE rating uses. However, though these numbers may look a little optimistic compared to SAE numbers, you'll find the temperature and pressure corrections more in line with those found in a race car. This means if you pay attention to the engine's operating environment, it will be producing much the same numbers as the standard corrected numbers indicate. You'll have to pay attention to the car's induction system and the engine's coolant and charge temperature management. Do that and the numbers quoted within these pages will be yours. Fail to do this and your motor could be easily losing 10 percent of its torque throughout the RPM range.

Although the dyno is the best place to get power figures, it's not the only place. You can get a good idea of how much horsepower your engine makes by testing at the drag strip. Dyno testing and setting up, either on the dyno or at the strip, will be discussed in several following chapters. We now have looked at the fundamental principles of developing horsepower from an engine and what horsepower is. Let's see how these fundamentals can be successfully applied to the small-block Chevrolet.

Here is an example of power being overstated. This car belonged to a friend of a friend and the 383 crate motor purchased for it was supposed to make some 475 hp. On this precisely calibrated Dynojet chassis dyno it made a little over 300 at the wheels. If it was truly a 475 hp engine it should have made about 390 to 400 hp at the wheels.

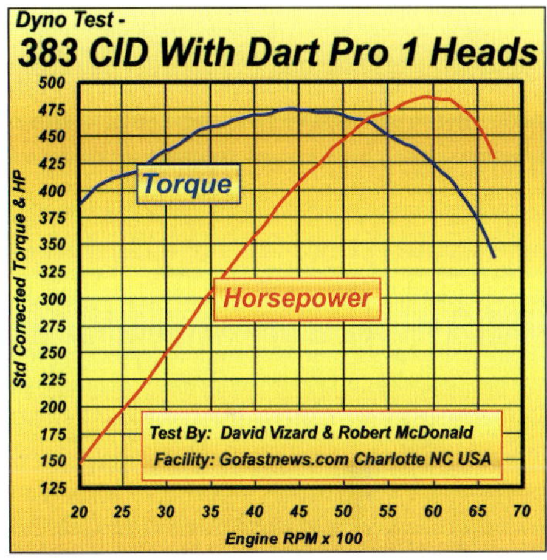

You will see a lot of power output graphs throughout this book. What you will not see is the graphs printed directly from the dyno readout. These are usually difficult to read and are not always appropriately scaled to show the differences between two tests. Instead, all the dyno figures are put into MS Excel and drawn as you see here to clearly show the information under consideration. For the most part, torque will be shown in blue and horsepower in red. Also note that all the figures are in Standard Corrected format. This means they will be about 4-percent higher than the SAE figures currently used by Detroit car manufactures, i.e., 100 SAE HP = 104 Std Corrected HP.

CYLINDER BLOCKS

As you may suspect, there's horsepower in blocks. However, it's amazing how often an inappropriate block choice is made for a high-performance motor. It might seem obvious but it's worth stating anyway: big bores and big cubic inches will always make more usable power than small bores and small cubic inches, even if the smaller engine is turned to proportionally higher RPM. This is especially so for a true street machine that's expected to perform and be civilized. The only factor that should prevent you from adopting the largest bore/displacement block possible is long-term reliability or money. If you have read and absorbed the seemingly obvious forgoing, you will be a step in front of at least 25 percent of those attempting to build a low-cost performance small-block Chevy. Having hopefully driven the "no substitute for cubes" point home, let's start with what to look for and what to avoid.

Block Logic

The first and most important rule to be observed is to never use whatever you already may have simply because you have it. Building anything of fewer cubes than you can afford (and a 350-cubic-inch motor is about the bottom line here)

The foundation of the motor is the block. Make sure that the build starts with a sound block, appropriately machined and prepped.

is an outright bad move. At the end of the day you will have paid more per horsepower than if a larger motor had been used as a starting point. There are four reasons for this and they are as follows:

1. Torque accelerates a vehicle and cubes deliver torque.
2. Achieving a decent compression ratio on a smaller-inch engine compromises the combustion chamber shape and consequently output—especially torque!
3. The valvetrain will have to deal with higher RPM on a smaller motor and high-quality components to handle the higher RPM will become more expensive.
4. Smaller bores compromise the valve

sizes that can be used. With a small-block Chevy, bigger is better to a greater extent than in most other engines. Always keep that in mind as you plan your motor.

What to Look For

Let's take a quick look at the small-block Chevy's family tree. Introduced in 1955 at 265 inches, it was expanded to 283 cubes in 1957. In the 1962 model year the Corvette came out with 327 cubes. This was the first of the 4-inch-bore motors and set the stage for what was to become the most common bore size among small-blocks. Getting 327 inches was not just a case of extra bore

If there is any doubt as to the thickness of the cylinder walls, a rough guide can be had by using a set of bow calipers to reach down into the water jacket as seen here. Next step is to measure across the adjusting screw and note the dimension.

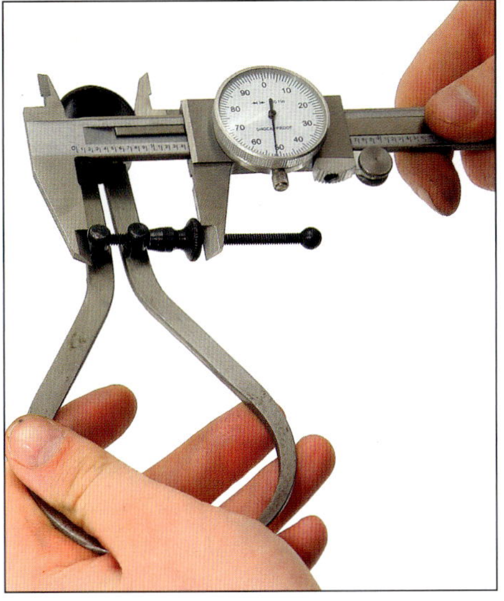

After removing the caliper from the block, the dimension seen while measuring into the block is re-established. Measuring the gap between the caliper ends will give a reasonably accurate dimension for the cylinder wall thickness at that point.

Seen here is the rear main cap on an early model block. The cap houses a two-piece, rear main bearing oil seal.

diameter. The stroke was also increased from 3.00 to 3.25 inches. This move necessitated larger crank counterweights and, as a result, the internal form of the crankcase was changed to give added clearance. Good as the 327 was, the 1967 model year was when the 350 was introduced. Not only was this block able to accommodate a longer stroke, but cranks for this displacement were also beefed up with larger journals.

The following year, 1968, saw the introduction of the 307 engine. It had the bore of the 283 and the stroke of the 327. It has absolutely no place in a performance program. From 1967 to 1969 the 4-inch-bore, 3-inch-stroke 302 was produced. Essentially it was produced just so Chevrolet had an engine eligible for high-profile TransAm racing. The next major milestone in terms of performance was the introduction in 1970 of the 400-cubic-inch block. This deviated from the previous standard format in many respects. First, it was the only small-block that Chevrolet made with 4.125-inch

bores as opposed to the by-then-common 4-inch bore.

To achieve the 4.125-inch bores, some internal casting changes were done. This block usually is recognizable by the fact that most of them have three freeze plugs in the outside of the block rather than two. To accommodate the larger bore at the same bore centers, it was necessary to delete the intercylinder water jacketing common to all other blocks except the Bowtie blocks. This style of block is commonly known as a "Siamese bore" block. Along with the increased bore size, the 400 block was made to accommodate a crankshaft that had larger main bearings. It should be noted here that throughout the life of the small block, GM was very generous with its selection of crank, rod, and main journal size. As a result we find that aftermarket cranks made with quality modern materials are more than strong enough, even at the smaller sizes to deal with strokes up to 3.750 inches in cast-steel form and 4 inches in forged/billet form.

In 1987, Chevrolet made some serious updates on the small-block. Essentially the blocks from here on out were equipped from the factory with hydraulic roller cams instead of flat-tappet cams. Basing your build on a later roller block has some distinct budget power advantages because, with a few lifter mods, you can end up with a highly functional roller valvetrain for less money than rebuilding a flat-tappet valvetrain (more on that in Chapter 7). Along with the change to a hydraulic roller valvetrain, the factory also changed the way the rear main oil seal is made. Here the rear main bearing and the oil seal diameter were changed and the block redesigned accordingly to utilize a one-piece rear main seal. The only real concern we need have here is the interchangeability between cranks and blocks from the earlier two-piece era to the later one-piece era.

Should you have an earlier high dollar crank, a rear main seal conversion kit can be had to allow it to be used in the later block. However, the kit is more money than a high-performance cast-steel crank so, unless that early crank is a real high performance piece, making the conversion is only just cost effective.

The one-piece rear main seal-style block is easily identified by the aluminum seal housing bolted to the back of the block.

In addition to the hydraulic roller cam and one-piece rear main seal there were numerous other design changes. These included a change of damper style and flex plate/flywheel changes to reflect whether an internally or externally balanced crank design was used. This aspect is something that you need to pay close

Preferred late-model block casting numbers 393638 (1987–1995) and xxx880 (1996–2002)

These two blocks are the most common and well supported by the speed equipment industry.

The number 327xxx block (1992–1997) is not usually used because of the reverse cooling. This was used in Corvettes with a 4-bolt main and Camaros/Firebirds with 2-bolt mains. All used aluminum heads. The number-327xxx block was also used in the Caprice but with iron heads. These blocks were built in far fewer numbers.

The least expensive rear main seal conversion kit for a late-model (post-1987) block can be had from AAEQ in Chicago. These conversion kits, shown here with the alignment tool, come in two styles—one for the small pan seal and one for the large pan seal.

attention to so that the final parts assembly is compatible; otherwise, the engine will be a long way out of balance. We will deal with this important subject in detail in Chapter 4.

In 1987 the heads also saw a great change in design. This was the year of the much-vaunted "vortex" heads. In terms of street performance over a very wide RPM range, these were some of the best production 2-valve-per-cylinder heads ever produced not just by GM, but by anyone. As far as interchangeability from early to late blocks is concerned, this was 100 percent so long as the relevant intake manifold and valve covers were used.

Apart from the principle players so far mentioned, Chevrolet also brought out a 262-inch motor, a 305 (which is now one of the most common), and a 267. You should not be considering any of these blocks as a basis for a high-performance build, especially when a tight budget is involved. Remember, and never lose sight of, the fact that cubes are the easiest and cheapest route to strong street performance.

Since its introduction in 1955 as a 265-ci unit, the small-block Chevy has grown somewhat in size. The most popular and prolific block, the 350, and largest stock displacement (and now relatively rare) 400-ci blocks, are the only ones you should be concerned with. These two displacement sizes differ in both the bore

and stroke. The 350 has a 4-inch bore with a 3.48-inch stroke. Not only is this a well-proven configuration in terms of reliability but also 350 cores at the wrecking yard are, figuratively speaking, nearly a dime a dozen.

Chevy 400 blocks are out there, but as the years go by their numbers diminish significantly, by attrition and hot rod usage. They are nice to have because of the extra cubes, but can be problematic. They can be a little more difficult to cool effectively at higher outputs and many have bore walls too thin for a performance motor. Knowing whether or not the block you may be eyeing fits that format can be a concern. I have had 400 blocks that have walls so thin that, at plus-0.030 overbore, the bores flexed as the hone went through. On the other hand, I've had 400 blocks that have had cylinder walls so thick (0.300 plus) they could have been the models for the original Bowtie block.

With 0.300 cylinder wall thickness and a suitable head gasket, a 0.125 overbore is possible but 0.060 is usually all that is convenient. Such blocks are a great find. Not only can the 4.185-inch bore go to give 412 torque-developing inches, but also this big bore helps unshroud the valves (see Chapter 6) a little more for better breathing.

Rod failure in any of the bores causes a chunk of the wall of the bore, at the position indicated, to be knocked out. This usually does not affect usage, but you should check for a crack into the water jacket.

Block Selection

The first step when selecting a block is to see whether the particular block is compatible with the vehicle the engine is going in, as well as with other parts you may want to use. If you're building a late-model motor, you may want to stay with a one-piece, rear main seal-style block. In that case, you'll be looking for a block made after 1987. It's also necessary to select a block that has the relevant external fittings for your vehicle's needs. A prime instance here is the threaded hole for a manual transmission clutch linkage. Also, the location of the dipstick varies on models after about 1976.

When you're sure you have a block that's compatible, your next target will be to determine main bearing bolt status. The question here is, will your application work with the cheaper 2-bolt block or does it demand a 4-bolt?

When it comes to performance capability, 2-bolt blocks have been somewhat maligned. A 2-bolt block can live with a lot more horsepower than usually accredited. When the budget is really tight take heart in the fact a 2-bolt block can, for drag racing applications, handle in excess of 500 hp with nothing more than a change from regular main bolts to ARP main studs. If you can get a 4-bolt block so much the better, but these are in high demand and their asking price reflects that. In many places you'll have to pay $50 to as much as $100 more for the privilege of having those extra main-bearing bolts.

Casting Material

Assuming that we're dealing with a sound casting with no obvious flaws and it's of 350 inches or more, there's really no such thing as a bad Chevy block. There's just good or better. The most desirable among the pre-1987 blocks, are the "high-nickel" ones. These blocks are cast with a two-percent nickel-alloy cast iron that hardens and gives significantly

longer bore life. The best blocks to get are those that have a number 10 or 20 in the area normally under the timing chain cover. These have one-percent tin and two-percent nickel. The tin is used to help the melt flow better into the casting mold.

These 10/20 blocks are the least prone to cracking. Also, because they pour more easily, they have the fewest problems with hot spots caused by porous metal. If you find a block that only has one number that's either 10 or 20, this means it has no additional tin but does have one- or two-percent nickel. If there's no number on the block it has neither additional tin nor nickel. These blocks can be expected to wear faster than others, but, with modern oils, are still good for 100,000-plus miles.

Locating the Right Block

There are two ways to locate a suitable block: one will keep your hands relatively clean, the other most certainly will not. The second way will get you more basic parts and save money, but will involve a lot of hassle. We'll deal with this method first.

Going the second route entails buying a core, usually from a wrecking yard or similar source. You'll have to identify this core by casting number. While you're at it, you may as well identify the cylinder heads. The chances are whatever heads come with this motor won't be the ones you want, especially if it's a smog motor. Essentially what you're looking for is a 350 or 400. Unless you are targeting a long block equipped with Vortec heads which you intend to use (rather than aftermarket heads) you should look to buy a complete block less heads. This will allow you to measure the bores and maybe see if there's internal damage in the engine. With the older engines this is a reality you need to consider. If you can afford a little more (usually about $50 to $100 more) then a late model (post-1987) engine has some intrinsic advantages.

These engines came with fuel injection and the far superior fuel metering on the cold start cycle brought about a huge reduction in bore wear. The advantage here is that if the engine has done no more than 120,000 miles and has had regular servicing, you can, in most instances, get away with a cleanup honing job to restore the bore/piston situation to "as new" condition. Also these later model engines have much better rods and the stock pistons are also good because they have a very good low drag ring package. If you intend to buy a later model engine buy it with the heads on because this protects the bores from the elements. If someone removes the heads then the condition of the bores is on a downhill path right there. It won't take long before rust dictates that a re-bore is a necessity rather than an option.

Once you've identified the motor you need, pull off the crank damper, timing chain cover, and timing chain to check the tin and nickel alloying content of the block. Block casting accuracy for the pre-1987 blocks was far from a precise process. With the update in 1987 came more precision castings, so that's another advantage to the later hydraulic roller block.

If you are looking at an earlier block you provisionally need to establish if the block has undergone no more than minimal core shift during the casting process. This can be roughly gauged by how even the metal thickness is around the lifter bores and the cam bearing bores at either end of the block. The more even it is, the stronger the block is likely to be.

If at this point the block looks worthwhile, pull the pan off if you haven't already done so and inspect for any obvious internal block damage. The most common damage is pieces missing from the bottom of the bores due to a rod failure. If the block passes this inspection and fits your main bearing bolt and budget requirements, buy it. Before leaving the subject of 2- or 4-bolt mains, I should point out it's entirely possible to convert.

Sourcing a Core Block

Don't have time to go to the wrecking yard? I am very much in that situation. Demands on my time start at about 5am and don't stop until about 8 or 9pm. So here is how I handle my core motor situation. I pick up the phone and call AAEQ. This is a big recycling company with thousands of core motors on hand at any one time. I tell them what I want—they quote me a top and bottom price depending on condition and parts already on the core. When they have what I want they ship it by truck to my door. All I have to worry about is a cart to unload it on. In addition to cores they have a range of low-cost parts such as performance heads, pans and valve and timing covers etc that are really smart and cheap. Hit their website at aaeq.net to get an idea of your options here.

You can either locate some mains from an otherwise scrap 4-bolt block or convert a 2- to 4-bolt mains by using aftermarket steel caps. Either way, they'll need fitting and line boring. In terms of outright strength, aftermarket angle bolt caps are the best option because they enhance block integrity. Unfortunately, even though some of the aftermarket mains caps are very reasonably priced (PRW has a great deal here), installation can be a little costly and should be considered outside the realms of a tight budget build but not so a budget extending to the $5,000 mark.

Teardown and Cleaning

Once you're satisfied with the block assembly, scrape off most of the dirt on it before you cart it away. On the way home, stop at a do-it-yourself car wash. Using Gunk degreaser and a bristle brush, clean the motor and blast it with hot, high-pressure water. When you get it back to your garage, mount it on an engine stand and ready your tools for the teardown. Don't try the rebuild without an engine stand; they cost as little as $45 for a sturdy-enough stand.

As the motor is stripped, you should look for signs of potential problems. Your first job is to pull the manifold if one is still in place. If it's a two-barrel manifold, junk it. If it's a four-barrel, put it aside. You can make a decision as to manifold choice after you've read Chapter 8. The next item to be stripped is the valvetrain assembly. If you're working on a really tight budget, check every rocker tip and the ball pivot area for wear. If they look reusable, wire every rocker to its ball pivot and place it with its pushrod (and later the lifter) in a plastic bag so they remain as a group thus allowing them a greater chance of being reused. Remove the heads and set them to one side for later evaluation.

Once you are absolutely sure it has been drained of oil, turn the motor over and remove the pan. Next remove the oil pump pickup and cover and take a look inside the pump. If the gears look to be in good condition, you probably won't need to buy a new pump. If the pump is good, replace the cover then remove the complete pump assembly and bag it.

Now remove the major parts from the bottom end. First remove each complete rod assembly one at a time, numbering them as you go. If the motor was serviced frequently, the crank probably won't need a regrind. This means you need to take special care not to damage the journals with the rod bolts as the rods are removed. To achieve this use a short piece of rubber hose over the rod bolt threads during disassembly.

As you remove the rod caps, inspect them and the crank journal closely. If the bearing material has worn away completely and it's steel-on-steel, or if the rod has any discoloration due to temperature, you'll need to have both parts checked before you reuse them. This means crack testing the crank and having the rod's big end bores checked for size and roundness. The ideal situation is for the crank journals to have zero to minimal wear and both the main and rod journal bearings in good condition. Not that you'll reuse the bearings, but good condition here indicates a cared-for engine.

Removing the rod and piston assembly from the last part of the bore often can present problems. Wear and/or carbon buildup will mean the top ring will encounter a step at the top of the bore. You can simply use enough force to drive the pistons out, which is okay if you intend to buy new pistons.

However, the later, post-1987 pistons are often fit for reuse and will deal with 425 hp without any undue problems. Because of this I recommend a little patience; the way to do the job is to scrape all the carbon deposits from the top of each bore before attempting piston removal. If the piston fails to come out relatively easily after this, you needed new ones anyway because the size of the step indicates the need for a re-bore. A good disassembly tip here is to have a box with a lot of crumpled paper in it right under the piston because, when the rings clear the top of the bore, the piston/rod assembly will drop like a rock. If not caught by a helper or given a soft landing you could damage the ring land to the extent the piston will be rendered unusable.

Regardless of status on removal, inspect each piston/rod assembly as it's removed from the block. An even skirt wear pattern across the pin axis indicates the rod is straight. If it's biased to one side, the rod is probably bent. If it is, junk it. The days of straightening rods are gone. It takes a fixture to check straightness and the cost of doing so is not worth the effort because new rods of far superior strength are available at a very reasonable cost. Do not reuse any rod that may exhibit even the slightest sign of being out of true because even a minor error here can cause a large increase in engine friction, which cuts horsepower substantially. So that you know the importance of straight rods, know that just one bent rod can suck up 20 hp!

Once the block is stripped, it needs to be checked for possible cracks and wear. For cracks, a magnifying glass is your first recourse. For wear on the bores, figure that if you can feel a ridge with a fingernail, then, at the very least, your block will need a cleanup hone.

If you're working on a really tight budget and chose your block carefully, there's a good chance that you can get away without re-boring the block. On many occasions I had low-wear blocks honed up to 5 thousandths to remove the ridge. I would then take the pistons that came out of those bores, knurl the skirts, and subsequently file-fit the pistons to the bores.

Granted, this sounds like a far-from-ideal way of building a high-performance engine, but we're talking budget. This penny-pinching route is far more viable than might be expected. I have knurled pistons with 100,000+ miles on them and, after a subsequent 80,000 street miles and an occasional pass down the drag strip, only exhibited 0.0003 inch of skirt wear. Assuming a maximum wear allowance of 0.001 or so, this indicated that the piston skirts were good for another couple hundred thousand miles. If you acquired a late-model block, then some very viable cost saving options are open to you. Since the wear factor is usually very low we find that the bores and the pistons have enough service life left in them to be totally re-usable for a mild to mid-level performance engine. The factory spec of piston skirt to wall clearances are intended to produce a quiet, slap-free piston operation. This is not the best for power as it incurs added piston friction. What this means is we can hone usually about 0.002 inches out of the block and get optimal power piston to wall clearances. There is more to the piston situation yet but we will deal with that and about saving pistons in Chapter 4.

OK, on with the block. The next move is to look at the bottom of the bores. These days many of the older small-blocks have already been through the rebuild mill at least once, and yours could have been rebuilt before. Check if the block has been rebored and/or linered. You're looking for a bore that's stock or (certainly no more than) near-zero-wear +0.030, for example. A 0.030-over block with wear of any consequence may not clean up at +0.040, and about 50 percent of blocks are a little too thin for a performance application at +0.060. Before I consider taking a block to +0.060, I sonic check all the bores to determine the viability of a 0.060 overbore. In my experience, about half the blocks will be fine and have walls thicker than the minimums I prefer. The later blocks seem more consistent in terms of cylinder wall thickness. This is important, because core shift may ultimately be the deciding factor between a viable 0.060-over and a no-go.

As mentioned earlier, check to see if at one time or another a rod has broken in the block. Evidence of this usually can be seen in the form of a chunk of metal cracked out from the bottom of the bore. If this has happened, it doesn't mean the block is now useless, since most blocks survive this catastrophe. What it does mean is you should have the block crack tested. Whereas a crack may not have worsened much during ordinary street use, we're going to be asking a lot more of it, so the block must be crack free.

What if you find your block has a liner; is that the end of it? No, I have blocks that had served me well and, for one reason or another, a part such as a wrist pin failed and gouged a cylinder. These were saved by linering. I have a block with a liner in each bank and it has survived many dyno pulls at around 600 hp. What I do not like to see is a block with two liners adjacent because I feel that can compromise the integrity of the block in the mains area.

At the end of the day, it really boils down to how thick the block walls are as well as correct liner installation. It isn't a difficult job, so for the most part a block with liners will suffice. If you find a cylinder bore has a large score in it due to a wrist pin working loose, if it fits your requirements and every other aspect, investigate the cost of linering.

Block Strip

At this point you should have an idea of whether the block is going to be usable, so a trip to your machine shop is in your future. If you intend to do the work outlined in the following pages, it will involve quite a bit of labor. You may feel it's worthwhile taking the block to your machine shop to have it sonic tested to be sure it's worth the effort. Not all machine shops have a sonic tester, so call around. For more info here, read the upcoming "Professional Cleaning and Checking" section.

If you want to save a few dollars, make your next job stripping the block of its various plugs. Removing any block plug is a matter of technique. The freeze plugs on the sides and back end of the block are easy to remove. All you need is a large round punch, a big hammer, and a pair of channel locks. The technique is to punch the freeze plug into the block so one side rotates in while the other comes out. This allows grabbing hold of the outer edge and, using the rounded side of the channel locks against the block, prying it out. Once you get the hang of this it usually takes about 60 seconds per plug.

The oil-galley plugs, especially those at the back of the block, can present a problem. There are two ways to remove them. Drill them out or use an oxy/acetylene torch to persuade them out the easy way. I prefer the torch method. The technique is to heat the plugs until red hot, then douse them in engine oil. Heating the plug causes it to expand and soften. The block, being cooler around it, squeezes the plug in; so, on cooling, the plug becomes smaller. When the oil hits the red-hot plug it becomes thinner and, by capillary action, is drawn into the threads.

After this, you'll find a quarter square socket will simply wind them out. When you're stripping the block of plugs, be sure to get the one in the oil galley under the rear main bearing cap, because there's almost always junk behind it.

The last job is to punch the head-locating dowels into the water jacket. Sometimes these aren't through holes. If they don't immediately go through into the water jacket, have your machine shop remove them—most have a special tool. Last, check that all of the critical head and main bolt threads are in good order. You will, for certain, have to go through all the bolt holes with a tap. This will achieve two things: First, clearing out the threads will establish their fitness for further use; and, second, it will ready them for further use. Take note here that only clean threads in good condition will establish the correct bolt preload to hold a head gasket when the bolts are torqued up.

Pre-Machining Details

Some of the photos scattered around these pages show detailing that can be done to improve both the appearance and the oil drain-back capability of your block. This should be carried out before any machine shop block cleaning.

Detailing also has the potential to produce a little more horsepower as well as enhance reliability. This is where your die

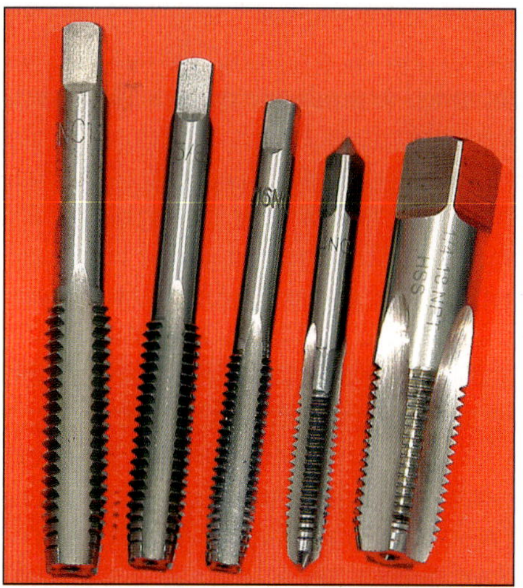

The taps most needed are (left to right) 7/16, 3/8, 5/16, 1/4 coarse, and 1/4 pipe.

There are plenty of sharp corners that can stand removing from the outside of the block.

grinder comes in handy again. Detailing is time consuming, but paying a machine shop, even if they were familiar with the job, would cost you several hundred dollars. This means you can look on the expense of your grinder as already redeemed.

The place to start is the rear main caps. Take a look at the nearby photo and you'll see how the oil groove around the oil pump securing bolt is: A) cleaned up; and, B) made deeper if it's too shallow. This groove is then blended into the hole that leads down through the main cap. The idea is to "flow" the oil passages so that the amount of oil pressure lost, as the oil progresses from the pump to the

To make it a little easier for the pump to do its job, clean up the area around the pump securing bolt on the rear main and blend it into the discharge hole leading into the block.

In this view of the filter mount area you can see how the passage into the filter and the exits have been streamlined to facilitate flow and cut pressure drop to the bearings.

bearings, is reduced. By doing this we reduced or eliminated the need for a power-consuming, high-pressure, high-volume pump.

Next, look at the entrance and exit holes in the oil filter location housing. Blend these holes into the casting. There's also some modification you can do to the oil pump, but we'll talk about that in Chapter 5.

Oil Return Holes

We don't want any more oil moving around inside the crankcase than necessary. On the pre-1987 blocks, which all used flat-tappet cams, there are oil return holes up the center of the lifter valley. Oil dropping through these holes in the lower part of the crankcase becomes caught up in the rotating assembly. It's better to have the oil return at either end of the block. At the back of the block, oil can return beside the distributor drive.

This hole should be ground and cleaned out. There are two good reasons for doing this. First, we want the oil to return as quickly as possible to the pan, and secondly, these areas at the back of the block are a casting sand trap. If any sand is left in the casting, this is almost certainly where it will be. Grinding the block smoothly in this area will eliminate that.

The other area to work on is at the front of the block. The two large holes near the lifter valley floor should be ground so that the oil can drain down and out the front of the block via the cam timing cover. If you intend to use regular chilled-iron, flat face lifters, it's not a good idea to starve the camshaft of oil, because a regular-iron flat-tappet lives or dies because a result of its lubrication quality.

Check the areas that drain the oil back to the crankcase on your stock block and you will see that the reworking shown here improves the drainback considerably.

In this shot the brass standpipes can be seen. You need a 1/4 pipe tap and some double-ended 1/4 pipe fittings from your local hardware store to get this job done.

In the case of a flat-tappet cam, the value of plugging the holes up the center of the lifter valley is moot. However, if a roller cam or the hard-faced flat lifters will be used, then oiling the camshaft/lifter interface is far less critical. This means you can cut excessive drainback from this area by installing standoff pipes or plugs in the lifter valley holes. I prefer the standoff pipes since they allow the crankcase and the valley area to breathe freely without allowing a significant amount of oil to pass through.

The next job, although time consuming and somewhat costly, is important. Tap all the head and main bearing bolt hole threads with a new, high-quality/high-speed steel tap (don't use a cheap one in a plastic box from an auto shop unless it is absolutely new). These block threads must be clean and in good condition, otherwise the required clamping forces will not be realized.

Professional Cleaning and Checking

Having come this far, it's time to take the block to your local machine shop. The machinist's first move is to clean the block. One of two cleaning methods will be employed: hot tanking or the hot-bake/shot-peening cabinet technique. The second method is by far the best because it leaves castings looking better than new.

Inspect the block prior to machining. If your machinist has a sonic tester, have the bore wall thickness checked. As I have already indicated these can vary considerably, both in evenness and overall wall thickness, on 350 blocks and—even more so—on 400s. In the days before sonic testers, most 350 blocks for performance use were normally bored only 0.030 over. However, since there are a lot of blocks that already have gone through a standard and 30 overbore, it's worth investigating the practicality, if needed, of a larger bore yet.

There are many 350 blocks out there that will bore 60 over and still leave more material than many at only 30 over. If you have such a block, you have the option of going up to 60 over. Personally, because I'm a nitrous user, I don't like to see bore wall thickness drop much below 0.150 and I much prefer to see them at 0.180 or more.

If your machinist doesn't have a sonic tester, it's not the end of the world. At one time, high-performance piston availability was limited to between +0.030 and +0.060. However, the situation in the last few years has changed and budget high-performance pistons can be had in almost any convenient oversize required. If you have a motor that's worn out at +0.030, see if it will clean up at +0.040. Your machinist will advise you. If it won't clean up at +0.040 and you have to go +0.060, be aware that it's best at this point to get the block sonic tested unless you anticipate an output less than about 350 hp.

Moving Bores

Because of core shift during the casting process, the bores are rarely in the middle of the surrounding core of metal. It can be frustrating to find that most of the cylinder walls are thick enough to be overbored, but just one or two cylinders are thin on one side to the point you're reluctant to use the block. This apparent problem may not be as big an issue as you may think. Because the pistons and rod assembly have a considerable amount of leeway sideways, it's possible to move the bores along the length of the block (the most frequent place they are thin) so that the cylinders can be bored more centrally in relation to the material surrounding them.

For instance, if a cylinder wall is 0.220 on one side and only 0.160 on the other, and the intent is to bore 0.030 bigger, it's practical to move the cylinder over as much as 0.025 if bore wear will allow. Such a move means most of the material comes off the thick side, leaving a spread, side to side, of 0.195 and 0.155 instead of 0.205 and 0.130. Whereas this process cannot be carried on indefinitely, it certainly is practical to move the bores as much as 0.025, which often will make the difference between using a block or not. Best of all, it allows bigger bores that would not have been possible without the aid of sonic testing.

The more highly loaded side of the bores is the major thrust face, which is where the greater wall thickness is more desirable. If you're looking at the front of the engine, the major thrust face is the left-hand side of the cylinders. If the cylinder wall is particularly thick on that side, then you do have another power-producing option. It doesn't provide a significant advantage on a low-compression engine, but on a high-compression engine it's well worth the effort.

This option involves offsetting the bores in the direction of the thrust face. This causes the rod to have greater leverage on the crankshaft immediately after top dead center (TDC). At TDC in a normal situation, the piston, pin, rod, and crank are all lying on a common center line so that no matter how much pressure there is on the piston, no torque is seen at the crank. The offset allows the crank to start delivering torque from the combustion pressure sooner in the cycle. Usually about as much as you can go here is 0.020 because of head gasket constraints.

The usual technique for achieving this effect is to have a piston with an off-set pin. However, this has disadvantages because most piston manufacturers don't like making offset Chevy pistons. Also, a piston with pin offset has more load on one side than the other. It gets cocked in the bore and increases bore wall friction. This usually isn't a disadvantage until the piston offset is in the region of 0.060 or more. However, when you go to 0.020 offset in the block (if the material is there to do it), there's no disadvantage.

Machining Operations

Whether you're buying a block already done or taking yours along, it's time to discuss machining operations. Because machining costs money, it's important to get the most for your dollar. Even the tightest budget should allow sufficient money to have the bores honed in some manner to clean them up. Let us first look at what is potentially your cheapest option. If you spent a little more

Because bore wear is often low to negligible on late-model fuel injected motors, we have often salvaged them for further use with nothing other than a glaze buster like this from NAPA or most other pro parts stores.

money and bought a later hydraulic roller block (from the 1987 model year on) there is a very real chance you will end up saving money overall. Because the bore wear on these fuel injected engines can be so much lower, there is better than a one-in-three possibility that you can simply "glaze bust" the bores and re-ring them up to near new condition. Even if you don't possess a glaze buster, one can be purchased from, say, AutoZone or NAPA for usually less than $28. Use a decent-size electric drill on slow speed and a lot of WD-40 as a lube and you can do a good facsimile of a re-hone in about two minutes per bore. When doing the glaze-busting op, be careful not to go through the bores too far because this will run the glaze buster into the main caps and break the stones. If you can afford to spend the money on a proper cylinder hone job then so much the better. Figure it will run about $100 or so.

The quality of the honing job is important. Honing heads with both stones and special brushes was a 1990s innovation for a better cylinder-wall and less-grit-laden finish. These honing heads have stones paired with stiff nylon brushes to follow the stone up, and in the process reduce the amount of honing

For a high-performance engine, the cylinder boring should be done on a machine such as this one at T&L. For positional accuracy, the bores are referenced from the main bearings.

debris pressed into the bore surface. This type of honing head is creating longer-lasting bores, which undergo less break-in wear. After break-in they provide better sealing for longer periods. Find a machine shop that uses this type of honing head.

These days, any decent, high-performance engine is finally honed with a dummy cylinder head called a "deck plate" on top of the block. This induces the stresses that cylinder head installation incurs. This operation produces rounder, more gas-tight bores when the engine is assembled. Honing this way is more expensive than the non-deck-plate honing as done on the factory production line.

The deck-plate method is the better way to go but if your budget won't allow that, the best plan (if basic honing is all the budget will stand) is to use regular soft iron piston rings. These rings, depending on the piston ring groove style, can range from dirt-cheap to relatively inexpensive. If the cylinder bore is prepared slightly on the rougher side, as is recommended for this type of ring, they'll quickly bed in and—to a certain degree—compensate for the fact that the bore isn't as round as it could be. Before opting to go this route, read Chapter 4, because there are some negatives involved.

If you go the deck-plate route, the extra quality of such a bore is partially wasted if you don't use a higher-quality, longer-life ring. This almost certainly means spending money for moly-faced top rings that are tougher and will last much longer. Normally, if you budgeted for a deck-plate honing job, it is as part of a re-bore and hone deal and new pistons will figure into the equation. However, if your late-model motor was really good in the bore department, then a deck-plate hone job to take out a couple of thousandths (to bring the skirt-to-piston clearance at the pin center to 0.004 inch) will mean an ideal power clearance for the stock pistons.

Your machine shop may suggest align honing the main bearings. My com-ment here is to forget it. It's an expense that returns little if anything in the majority of cases. Align honing subsequently can be done in the unlikely event the crank doesn't turn freely during assembly. A lot of engine builders will argue the point that friction must be reduced to a minimum.

My sentiments exactly, but align honing probably won't do this. Here's the logic behind this statement. A small-block Chevy making 700 hp can twist from corner to corner as much as 1/16 inch. Therefore, under running conditions, the bearings aren't close to being in line. Even if we were to have them exactly in line, it's unlikely that the engine would make significantly more horsepower. Why? Because the crank is riding in a bed of oil. The crankshaft is bending. It's able to absorb this bending. No matter how much force there may be, the potential loss in torque that occurs is due to the force times the radius at which the friction is acting. If the coefficient of friction is zero, then no matter how large the bending forces involved, there would be zero torque loss because any number multiplied by zero is still zero.

We don't have zero friction at the bearings, but it's low. The oil film between the crank and its bearings is very

stiff. Even if the bearings are out of line by a considerable amount, there will be no measurable difference in the power seen at the flywheel. Testing with deliberately distorted blocks and cranks supports this theory. The bottom line is, check main bearing alignment by installing just the bearings (no seal) lubed with very thin oil and torque up the caps. If the crank spins freely, you are in business and more than $100 to the good.

A machining operation you should definitely try to budget for is "block decking." A small-block Chevy has a deck height of 9.025 inches stock. Just about every piston made ends up 0.025 down the bore, which compromises desirable quench action and, on 350s at least, access to higher compression ratios. Decking the block to 9.000 inches for a Fel-Pro 0.038 gasket or 9.010 for a steel shim is worth 8 to 10 hp and reduces octane requirements.

If you want to do a trial build, first have the block decked to give 0.028 clearance between head and piston when a close-fitting piston is used. Some of my more wealthy/adventurous engine-building friends have run quench clearances down to 0.022. I'd advise against that unless you're prepared to sacrifice parts by experimenting. By making the quench area work better, we move the engine further away from detonation while increasing compression ratio. A good quench action can allow as much as one extra ratio to be used.

The second point to make is that on smaller-inch motors there's a problem getting a high compression because of the extra combustion cavity volume residing in the block. This volume is already damaging to power because it limits quench action and reduces compression ratio. By decking the block, we reduce the capacity in the top of the block by some 5cc.

Another reason for decking is that the deck surface between the two center cylinders directly beneath the exhaust

Using the main bearings as a reference, the block decks are squared and cut parallel to the mains at a 9.000-inch deck height.

The final operation on a serious effort block should be deck-plate honing done as shown here.

Deburr the lifter bores first with a de-burring tool or needle file, and then finish with a ball hone as seen here.

The three front oil galleys need to be tapped for a 1/4 pipe plug. Do not tap in too far or the adjoining galleys will be blocked.

Before you even think about installing the heads, make sure this oil galley plug is in and sealed up.

valves can sink. It's not uncommon to see this area low by 0.002 to as much as 0.004. Cheap steel-shim head gaskets work well on flat parts. If the block isn't flat it will be harder to hold a steel-shim head gasket and you may find yourself buying more expensive composition gaskets at a later date. Do it right now or do it over again later—take your pick!

Next to last is a small job that easily can be overlooked. The front three galley plugs are normally press-fit items. These should be tapped to 1/4-inch pipe threads. There's not much metal left to do this, so care must be taken during tapping.

The last op to do on the block is to chamfer all the sharp edges left from machining. That's machining done either by the factory or your machinist. Sharp edges left on the mains bores need to be filed off with a needle file. Also important are the possible sharp edges left on the lifter bores. The oven-bake/shot-peen block cleaning method can leave a burr around the edges of the lifter bores. These can shave off oil and cause the lifters to wear far quicker than would otherwise be the case. Be sure to put at least a 0.005-inch chamfer around the top and bottom of the lifter bores. Although this can be done with a file, it is very time consuming to get in on the crank side of the lifter bores. The best way to de-burr here is to get a small lifter bore "bottle brush" and go through the bores. It will remove metal from the corners way faster than from the lifter bore itself.

Acquiring Your Performance Block the Easy Way

Rather than spending time hunting in wrecking yards, you can acquire a block (with whatever machining you feel you can afford) ready to go. There are a lot of companies out there that will do blocks. In the past, when I have been in a tight spot to get a magazine engine done in time to meet a copy date, I have had Speed-O-Motive in West Covina, California, source a block and prep it for me. The block arrived by truck complete with a fully balanced rotating assembly. Allowing for this, Speed-O-Motive can do a short block for you from the most basic at $2,000 to about $3,500, like they did for me. For the most part the basic deal—block stripped, cleaned, bored, deck plate honed, decked, cam bearings installed and a coat of paint for protection—will get you going in a cost-effective fashion.

This will leave you to install the freeze, cam, and oil galley plugs. And it won't hurt if I remind you once again not to forget the plug that goes into the galley section under the rear main cap. Also, to avoid installing a head and finding a giant oil leak: do not overlook the screw-in plug at the rear right-hand side of the block deck face.

Post-Machining Detailing

At this point I am assuming you have sourced your own block and had the machining done on it, or that you obtained a most-basically machined block from a company supplying blocks.

First, let's start by checking the top of the bores. These should have had a small

This block had a basic prep done on it, but still absorbed about four hours of work. Here it is being readied for a couple of coats of primer and a coat or two of engine enamel.

chamfer applied by the machine shop to facilitate the installation of the piston assembly into the bores. Without a chamfer here, the rings will almost certainly hang up at every attempt to install the piston into the block. This chamfer should be neither too big nor too small. If it is too big, the rings, especially the narrow ones like the oil ring rails, can pop out from under the ring compressor and potentially get damaged during assembly, or even prevent assembly. If the chamfer is too small, assembly will be difficult. However, too small a chamfer is rare but the point I am leading to here is that the machine shop should have chamfered the tops of the bores. If they have not, don't go to that shop again because this is a failure on their part.

OK, so much for making a big deal about bore chamfering. Now let's cover stuff the machine shop won't have done unless you purchased a high-end block.

First, using about a 3/4-inch diameter chamfer bit in a slow drill, chamfer all the bolt holes in the deck face. This is especially important if you have had the block decked. With this done, do a final thread clean with a good tap on the head- and main-bearing bolt holes.

Let's now turn our attention to the bores. Honing has a nasty habit of impregnating the bore walls with honing grit. If your machine shop uses the brushed honing head mentioned earlier, this problem will be considerably reduced. Honing grit is highly abrasive and will cause the rings to wear rapidly unless it is removed. Honing grit is not the easiest to remove. Typical cleaning solvents do not readily remove it. The recommended way to go here is to use a stiff-bristle brush, hot water, and dishwashing soap and scrub the bores. This scrubbing with hot soapy water should be considered essential. Once done, this can be followed up with a little power-producing trick that I have employed over the years. Once you have scrubbed the bores, spray them down with Gunk. Then, using a Scotch-Brite scouring pad, rub the bores with a steep, almost-vertical spiral action until they feel as slippery as ice. This takes only two or three minutes per bore. Next, using more Gunk, loosen any dried deposits, and then hose down the block thoroughly with water. Blow off the block with an air line and hit all the machined surfaces with WD-40.

It's now time to check that the bores are really clean. Take a white paper towel, spray it with a little WD-40, and wipe down the bores. If they are truly clean there will be no discoloration on the white towel. If they are still not clean, you will find the paper towel has picked up the gray color from the honing stones.

What you see here is a decked and honed block, detailed out ready to go. The cam bearings need a special installing tool to get the job done. Either rent one from AutoZone, or better yet if you have never done this job, get them installed by a pro in a machine shop. Expect to pay in the region of $20 plus the cost of the bearings.

CRANKS, RODS AND PISTONS

Depending upon your approach, the crankshaft of your small-block Chevy can be either one of the cheapest or the most expensive power products you'll ever buy. Confused? Let me explain. Remember that nothing smaller than 350 ci is worth building. Unless cubes are restricted by class limits, you shouldn't use a crank of less than 3.48-inch stroke (stock 350). Correctly utilized, a crank with more stroke is cheap power, but if it breaks it will be expensive. The key is obtaining cubes reliably at a price reflecting a limited budget. With that in mind, let's look at our options.

Stock Cranks

If regularly serviced, Chevys are slow to wear. If, when torn down your core unit's crank is in good order, then, assuming a very tight budget, a crank journal polish is all that's needed. If it's less than near perfect, the crank journals can be reground to an undersize but the cost effectiveness of such other options is marginal. But first it should be understood that a measurably "out of round" or tapered journal in a high-output engine will survive full-throttle operation for only a short time. If the crank looks good and is sized up correctly as per the

Here is the Scat rotating assembly for a drag race 350 I built. As seen here this forged rotating assembly came balanced as a kit from Scat. In its track-ready form, the moderate budget engine these parts went into made just a couple of horsepower shy of 600.

chart on page 35, a regrind isn't necessary. It can be used confidently at power levels of 350 to 375 hp for early cranks and to about 425 hp for one-piece rear main seal cranks.

If the crank journals are out of tolerance, then you have some options. The cheapest is to have your existing crank reground. This will cost about $100 to $125 when done properly—ground with

as large a fillet radius between the journal and the crank cheeks as possible. In this respect, regrinding has advantages other than correcting bearing clearances. Many racers are concerned about undersized crank journals, reasoning that the crank must be weaker. But cranks don't break through the journals. A correctly reground crank can offer the racer a stronger part than a standard-size crank.

Factory Crankshafts: Cast or Forged?

You may have heard that a forged crank is the way to go. You heard correctly, but this should in no way imply that Chevrolet's cast cranks are second best—far from it. Chevrolet forged cranks are good, but the cast cranks are, for many applications, just as good, and in some instances, better. Properly prepped, a cast crank is fine in a circle-track motor up to about 400 or so horsepower, and I've seen them survive 200 passes down the strip in motors making 600 hp. Although the latter may be an extreme, and a certain amount of luck may be involved, it does show that they are relatively stout pieces with the later one-piece rear main seal cranks being even better.

Cracked forged cranks show up at the rate of about 10 to every cracked cast crank. This means that if you have a forged crank, it's imperative it be crack tested and discarded if any cracks are revealed. Although cast cranks have less outright strength than steel forgings, they have more internal damping. Unlike steel, they do not "ring" as easily and, as a result, if they're not stressed to their limit, the fatigue/wear life is as good or better. Although they may be made of a material with only 65 percent of the strength of a stock Chevy forged crank, they last and deal with stresses as if they were more like 80 percent of the strength. Also, cast iron has better wear properties than a non-heat-treated steel so it can last significantly longer. With a good damper, a cast crank is fine in a 425-hp street motor (where the duty cycle, full throttle to cruise, is mostly in cruise mode) and, remembering that most quoted horsepower numbers are inflated, 425 genuine horsepower makes any street car into a flyer.

Although it might not appear so, this is a stock crank. I reworked it with a die grinder and my lathe. Not only was the windage cut considerably, but also the moment of inertia due to a 10-pound-plus weight reduction. On the dyno all this work added up to a 12-hp gain at 7,500 rpm on the 300-rpm/sec acceleration range on a SuperFlow dyno. Cost involved: about $20 worth of grinding and finishing supplies.

Why? Because it can be ground leaving larger fillet radii in the corners between journals and crank cheeks. This is where 99 of 100 cranks break, and the bigger the radius, the smaller the chances of breakage. I've ground cranks with perfectly sized standard journals, but with what I considered too small a fillet radius, to 0.030 undersize for no other reason than to increase the fillet radius and hence reliability. For power levels above 300 hp and near continuous wide open throttle usage, check to see that there's a well-defined radius in every journal corner. If there isn't, your crank could break!

Now consider this: after possibly spending $125 on a crank regrind, what are you left with? Answer: Nothing more than a stock reground crank. Although it has acceptably sized journals, it is still of stock material and you may not know what sort of abuse it has already had and how much of its fatigue life is already used up.

The aftermarket in general and Scat in particular has much to offer the engine builder needing a rotating assembly at re-builder prices. Using one of Scat's stock replacement cranks means getting a new crank with larger fillet radii, and better material for about $50 to $75 more. Scat's entry-level crank is cast steel (not cast iron) and almost as tough and strong as a factory forged crank, which needs to be born in mind before deciding to re-grind the stock crank.

Regardless of the crank's regrind or new status, new bearings will be needed. This presents some cost-effective options we will get to later.

Considering the tight tolerances involved, the cost of grinding a crankshaft at about $100 or so is not that much, especially when done by a specialist using quality production equipment. Be aware, however, that most shops are more interested in turning over quantities for regular street use. To avoid grinding-wheel changes, high-volume shops will often grind as many cranks as possible using the same wheel.

When you ask for as large a fillet radius as can be had, you're seen not as a customer that will help their bank balance but as a potential pain in the neck. They may smile nicely and say what you want to hear, but you won't, in most cases, get anything different from whatever radius the wheel had on that day. If you must use a stock crank rather than a new-stock replacement crank, get your crank ground

Factory cast cranks are pretty decent pieces. Seen here is a stock cast late-model one-piece, rear main seal 350 crank.

Regardless of the crank source, all the bearing journals should be checked for accuracy.

at a performance-oriented machine shop and expect to pay accordingly. Such shops will take the time to change or dress a wheel to do the job right. They appreciate that you're knowledgeable if you're aware that a 0.030-under crank with big radii is stronger than a 0.010-under with small fillet radii.

The method used to clean a crank prior to machine work can also influence its reliability under high-stress usage. The oven/shot-peening process should be your method of choice. If the shot used is kept in good order rather than allowed to fragment, the process goes part way toward emulating a regular $125 shot peening process.

Choosing your machinist carefully provides you with a crank that will serve as well or better than a new crank in all respects but one. It may break due to fatigue from previous long, hard service, of which you're unaware. Let's see how best to guard against such a situation.

A Little Money, a Lot More Torque and Horsepower

It's a fact that 400-ci small-block Chevys wear out their bores far faster than their cranks. Built by Chevrolet in great numbers during the early and mid-1970s, the bulk of these engines became due for overhaul in the early 1980s. Bore wear plus cooling problems (fixable if you know how) did not endear these engines to many, so blocks with overly worn bores were scrapped by the thou-

sands. This left an equally large number of good 3.75-inch-stroke 400 cranks around, along with a lot of people wondering what to do with them. The solution to that problem didn't take long to figure out.

The 400 crank has a larger main bearing at 2.650 inches diameter than all other small-block Chevys. By grinding this down to the 2.450 diameter of the stock 350 crank, the 400 crank will pop right into a 350 block, adding some 28 cubes in the process.

Although the crank installation is total simplicity, installing the rods and pistons can bring about a number of minor problems. This, and the ways to avoid such when dealing with rods, is discussed later in this chapter.

Fast Mover

Because it makes such a strong street motor out of a regular 350, the 383 conversion became the most popular hardcore engine-rebuilding hop up move of the 1980s. Demand nearly outstripped the supply of cheap 400 crank cores. By the early 1990s, this left the would-be stroker 350 builder with a dilemma. By this time, all 400 cranks were old and had seen a lot of use, so the fatigue life was at the wrong end of the graph.

Finding a decent low-mileage crank to grind was becoming increasingly difficult, and prices for a good piece were escalating. Since so many 400 cranks had

been ground to fit in 350s, the used market had quite a few to offer. But buying one needs to be tempered by the thought that no one puts a stroker crank in Granny's shopping car. A used 350 stroker crank will be just that—used and, likely as not, used up.

The beauty of a 3.75-inch stroke in a 350 is that it allows improvements to be made in other areas and, if not compromised (as is often the case) by too short a rod, proves to be the way to go when hopping up a 350 on a budget. If you have or can get a 400 crank that's had an easy life and checks out crack free, you can grind it to suit your 350 block for about two and a half times more than the cost of a regular regrind. These days, the big problem with going the reground-400-crank route is that it's becoming next to impossible to locate a good core, and they're typically 50- to 75-percent more expensive than a 350 core. On top of this, you can never be sure of the crank's history.

The only factor working in your favor is that 400 motors were never used in a performance application. If you're going to build a shorter-cammed street torquer that will never see more than 5,500 rpm, a thoughtfully selected reground 400 crank is not likely to be a problem. However, the purpose of this book is to show how to build some real power, and a totally street-drivable, low-cost 383 can be built that really will put a reground stock 400 crank of unknown history to the test. Since building performance is the goal

CRANKSHAFT JOURNAL TOLERANCES	
End play	0.002-0.006
Rod Journals – All	
Standard	2.0990/2.1000
10 under	2.0890/2.0900
20 under	2.0790/2.0800
30 under	2.0690/2.0700
Main Journal – 350 bearings # 1-4	
Standard	2.4484/2.4493
10 under	2.4384/2.4393
20 under	2.4284/2.4293
30 under	2.4184/2.4193
Main Bearing # 5 (rear main)	
Standard	2.4479/2.4488
10 under	2.4379/2.4388
20 under	2.4279/2.4288
30 under	2.4179/2.4188
Main Journal – 400 bearings # 1-4	
Standard	2.6484/2.6493
10 under	2.6384/2.6393
20 under	2.6284/2.6293
30 under	2.6184/2.6193
Main Bearing # 5 (rear main)	
Standard	2.6479/2.6488
10 under	2.6379/2.6388
20 under	2.6279/2.6288
30 under	2.6179/2.6188
For race applications only, the crank can be up to 0.0005 under bottom limit for both rods and mains. For all journals maximum taper must not exceed 0.0005 and maximum runout 0.0005.	

this raises the question as to why we would even consider the use of a reground 400 crank here when Scat has such a good cast-steel crank that can often be had on sale from discount parts houses for as little as $199! My advice is to be sure of what you're doing; a crank breakage will be a huge negative impact to your finances.

Keeping that in mind, if you adopt the information within these pages, you could be looking at a 700-hp motor. I suggest only considering a reground crank if you're targeting no more than 450 hp or operating on the slimmest of budgets.

Keep in mind that if you spend money on a stout bottom end, you can always add more power-producing parts later when finances allow. For a little more than twice the cost of a regular regrind on a stock-stroke crank, you can step up to a considerably stronger, longer stroke, Scat cast-steel crank. Since most builds will be based on 350s, let's look at crank performance and economics for this situation.

Using the cost of a regrind as a yardstick, if you have to acquire a stock cast 350 core, your crank cost will double that of regrind on a core already owned. If you have a 400 crank, converting it will run 250 percent of a regular regrind. If you have to buy a 400 core and have it reground to suit a 350 block, it will cost about 375 to 400 percent of the cost of regrinding a stock 350 crank. Always remember that at the end of the day you still have a used crank. If you buy a Scat cast-steel crank to build a 383, it will have zero time on it, and from the right source it will cost about 300 percent of the cost of a straight 350 regrind. You can see how the economics are shaping up.

In 2009 dollars, the least expensive Scat crank, on sale, costs about $140 more than a stock crank regrind. This additional outlay in a 0.030-over block delivers 28 more cubic inches. If you build to the book (this book, that is), you are unlikely to see less than 1.1 ft-lbs of torque per cube with 1.2 to 1.25 being more common. This means the additional outlay of $140 delivers between 30 to 34 extra ft-lbs, with most of this increase showing up in the usable rev range. That works out to about $4.40 ft-lbs increase. To put that into perspective: not only do you get a new and stronger crank, but also the torque increase per ft-lbs is about 40 percent of the cost, of that delivered by a typical bolt-on blower installation. Assuming a moderate cam that allows peak power at 6,000 rpm, the extra inches provided by the crank are worth between 34 and 40 hp. If it's a race application, the advantages of a longer-stroke crank in a

350 don't stop there, especially if your budget means using production-type heads with a low set plug position.

Bigger displacement means the ability to achieve a higher compression ratio without compromising the chamber shape with excessively high piston crowns. Often it's possible to achieve the desired ratio with cheaper and lighter flat-top pistons. Having a better chamber in a 13:1 motor can alone account for a minimum of 10 additional horsepower. In terms of performance, the Scat 383 crank looks like a winner, but you may be asking: Just how good is a crank when it costs appreciably less than a new Chevy crank?

Back in the early 1990s I asked that question of a number of engine builders before I put one of these cranks into any of my motors. This was not because I thought the cranks might be of less strength than the stock cast crank, but because I've built low-cost engines that made four-figure horsepower numbers and broke stock parts. Scat has built a reputation based on quality over a period of more than 40 years, and the boss, Tom Lieb, isn't about to throw that away.

Unlike many cheap cranks, every Scat crank goes through an inspection routine to check bearing finishes and sizes similar to that of the $2,500 Scat Cup Car race cranks. As a supplier for professional Top Fuel, Cup Car, and TransAm among others, this company knows what it takes to make a crank live. I needed to know at what level I might expect failures. To this end I spoke to Bill Smith at Speedway Motors in Nebraska. At the time, Smith had used more than 3,000 of these cranks in budget/claimer race engines at power levels up to 550 hp and RPM levels to about 8,000, and had not seen one failure attributable to a Scat cast-steel crank. At the time of this writing, I've built more than 700 hp with a nitrous-injected motor on a Scat cast crank. I'm sure it is approaching the limit but I'm still no closer to knowing exactly where the line should be drawn.

Clean Sheet

One thing about redesigning something that's been around 40 years is that there's plenty of opportunity to document any problems. When Scat went to the drawing board, the intent wasn't to simply produce a copy crank, but to build a superior crank. This isn't too difficult if the price tag is open ended, but only half the street rodders out there have a budget for forged-steel or billet stroker cranks. In the low budget category, this limited Scat's production method to casting. Stock-cast Chevy cranks are made of Detroit's wonder metal—cast iron—but in a seriously upgraded form.

Chevrolet engineers have their act together when it comes to building a tough, cost-effective crank. This meant anyone following in their footsteps had their work cut out for them if improvements were to be made. Keeping costs reasonable and strength up, Scat opted to drop into that narrow window between cast iron and forged steel—cast steel. After heat treating, Scat's current 9000 series cranks are in excess of 100,000-psi yield with 6-percent elongation. Although a little more costly to cast, these cranks have a marked advantage in terms of strength and, equally important, are tougher and more ductile than the cast iron used for stock Chevy cranks.

The cast-iron alloy used for Chevy cranks produces, in a tensile test, some 3-percent elongation before failure. The Scat material, at 6 percent, doubles the stock Chevy figure and, although not proportional, is a good indication of just how much tougher it is.

There is, however, more to the Scat cast-steel crank than stronger material. It's designed from the outset to use a significantly more desirable 5.7- or 6.0-inch connecting rod. When Chevrolet opted to go shorter on the 400 rod, the result was a lighter rod/piston assembly than the 350. The 400 crank's counterweights reflect this because they're only intended

Seen here is a Scat cast-steel 9000 Series aero-counterweight 3¾-inch stroker crank. When used in a 350 this not only gets you a new stronger crank but also a big chunk of extra output.

to balance this lighter weight. Since it's a good move to install 5.7 or even the lighter variety of 6-inch rods, a crank designed from scratch should reflect these requirements. The Scat crank does this and a little more.

To appreciate the counterweight advantage of the Scat crank, you need to understand that a 400 crank is significantly short of counter-balance weight for the center four cylinders. As we turn up the RPM on our hopped-up stroker motors, the lack of sufficient counter-weight mass in the center of the engine results in an increase in center main bearing loads. This is already the highest loaded main bearing and we could be making it worse. The external balance weights of the 400 serve to smooth things out but do little to reduce internal main bearing loads. Indeed external balance mass puts additional loads on the front and rear main bearings. An external balance crank damper at the front also puts a severe bending load on the crank snout. This may be okay for an engine turning a maximum 5,000 rpm, but with the heads I describe in Chapter 6 you could be building a motor that will see as much as 7,700 rpm. If an external balance damper is used at this RPM the crank better be strong enough for the job, otherwise the snout will eventually part company from the rest of the crank!

In essence, external balance is a convenient production-line Band-Aid fix for insufficient counterweight in and around the center of the crank. So, as RPM increases, there's a tendency to bend the ends of the crank up and down. This, as has just been said, overloads the front, center, and rear main bearings at the expense of the two intermediate bearings. The problem is—and always has been—getting enough counterweight in the center of the crank. The only real cure for fixing a stock crank, and a costly one at that, is to install slugs of heavy (Mallory) metal in the counterweights.

But, when starting from a blank sheet of paper it is something of a different story. In practice, there's little room to maneuver with counterweight design within the confines of a conventional crank. Here Scat took the opportunity to include as much extra counterweight as possible, consistent with other production and usage constraints. Because the Scat crank has hollow big-end journals and slightly larger internal counter weighting, the main bearing loads are a little more evenly distributed than the stock Chevy crank. Also, if rods and pistons toward the lighter end of the scale are used, it balances internally, allowing you to use a stock 350 damper/flywheel/flexplate.

If you were in the market for a Scat Cup Car crank, this is what you would get. This super light crank, made from the very best materials available (4340 steel) and nitride hardened, will reliably turn 10,000 rpm at power levels of 850 hp and last for several 500-mile races. But, as you might expect, cranks like this are far from cheap. However, when buying at the other end of Scat's price range, what you get is a crank that incorporates as much of the high-end technology as is possible. This really shows up in the hollow journal and aero counterweight design.

Other Applications

So far, we've looked at the economies of Scat's cranks for a 350 stroked to 383. Obviously there are instances where the stroke or displacement is a race-mandated requirement. Excluding for the moment short-stroke exotics, this means using a 3.48-stroke crank. Unfortunately, finding a good used factory forging gets harder every year. The good news is that Scat's 3.48-stroke crank costs about 25-percent less than the 3.75-stroke.

Here's my view on the use of a new Scat cast-steel crank as a replacement for the stock factory forged 350 crank: If the factory forging is new or close to it and the price is right, then that's the right forging. Unfortunately, nearly new forged cranks don't crop up that often. If I had to choose between a used forged crank of unknown history needing a regrind and a new Scat cast crank, I'd go for the Scat crank because it's new and has proven reliability at the power levels we are mostly dealing with. It's also one-third the cost of a new Chevy forging.

If you're building a 400, play it safe. The extra cubes of this larger displacement make it a little easier to break an old and possibly tired Chevy casting. The extra money needed to buy a new Scat 400 crank will look insignificant when the stocker breaks. If you were fortunate enough to pick up a later 360-degree one-piece rear main seal block, Scat has stock- and long-strokes available for this application.

Buying a Forged or Billet Crank

While used forged and billet cranks are well within the budget constraints we're dealing with, new ones, with one or two exceptions, are hardly so. To help avoid catastrophic failure, only buy a used forged or billet performance crank on the basis that it's proven to be crack free. Remember, all race cranks have a hard life, and if it's used up before you get it you'll have nothing more than an expensive doorstop.

When looking for a used crank, you're likely to see a variety of brands. I haven't used every brand on the market. However, I have had experience with Callies, Crower, and Winberg for outright race cranks. All are top-notch manufacturers so if you find a used crack-free crank at a good price, it is

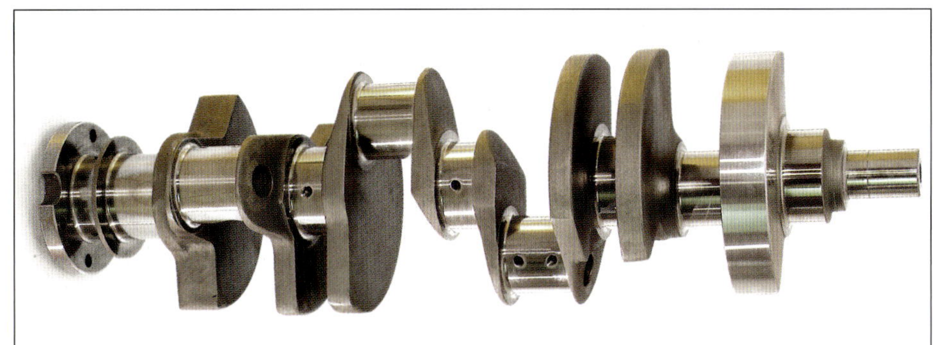

Scats standard forged cranks can be had in a variety of stroke lengths from 3.48 to 4.00 inches at a very economical price.

more than likely a good deal. The least expensive new forgings with assured quality come from the factory in non-heat-treated 1053 steel (110,000-psi ultimate tensile and 100,000-psi yield) and, once again, include Scat in the much stronger 4140. Both run about the same price, but the Scat cranks have the advantage of being better steel and are nitride hardened. The nitriding process, generally common to all quality cranks, forms a hard, wear-resistant case while producing a softer more ductile core, allowing greater flexure before breakage. Unlike cheaper (and usually off-shore produced) brands of forged cranks, Scat's budget forged cranks, as with its cast cranks, go through the same inspection given to its Winston Cup or Top Fuel cranks. These cranks are available in 3.48- to 4.00-inch-stroke lengths with the 2.45 main journal sized for the 350. To use them in a 400 block, spacer main bearings can be used.

Another brand of crank I have successfully used is produced by K1 Technologies. I have used their 4-inch-stroke crank to build a 408-inch engine out of a 350. This requires the right combination of block, crank, and rods; otherwise, cutting rod clearance into the block will hit water.

Bearings

Given the budget, I use Vandervell, Michigan/Cleavite 77, or Federal Mogul

King engine bearings are among the cheapest of reputable manufacturers and have shown good results in engines up to 480 hp.

Competition bearings. Admittedly they are more expensive, but still aren't that much money. If the output of the engine you're building is likely to be more than 450 hp, I recommend you do likewise. However, we do need to save where possible. For power levels less than this, bearings costing much less will work fine. Some of the "cheap" bearings on the market are low-cost partly because of the quantities sold and partly because they're cheaper to make. Certainly for power levels of 375 to maybe 400 hp, they're more than acceptable. In this instance you might want to look at King bearings. I have used quite a few of these in engines up to about 480 hp and the results look fine.

Strengthening Process

It's difficult to have a truly cost-effective component strengthening/hardening process at the low-cost end of the scale. All those I mention here are proven and are worth the money, but you need to consider the practicality of spending more for a better part rather than having a less-expensive part processed. Sometimes another $100 will get you a crank or rods that are heat-treated to the spec you are looking for and are of better material. The aforementioned economies apply big time to used stock parts but there are a few justifications for heat treating a new aftermarket cast crank such as produced by Scat. On occasions, I have had a budget for a little more than a cast steel crank, but not enough for an entry-level forged crank. What I have done here is to detail the entire cast crank with my die grinder. The crank shown above is just such an example. There is about 30 hours work in the grinding and detailing plus the time it took to do a coating job using the electric oven in the kitchen (wife not too impressed here!). Was all the effort worth it? Let's put it this way—all the detail work on every component that came under scrutiny such as the block, crank, oil

Using a die grinder, I detailed out this Scat 9000 Series crank then applied an oil shedding coating. Result: about 10 hp at 7,500 rpm!

pump, etc., totaled about 25 hp. This meant I could use a cam some 10 degrees shorter and have a more streetable motor that still made the top-end output of an engine with no detailing and a bigger cam. A back-to-back test showed the reworked crank work to be good for 7 to 9 hp at 6,800 rpm. This work plus the heat treating meant the crank was about 220 percent of its original cost but that is still only two-thirds the cost of a forged crank.

For heat-treating, there are two processes I have used to good effect over the years. These are proven cost effective and will make whatever part treated that much better.

First is ion nitriding (Nitron, Inc.). This is a relatively common process used in the improvement of mostly ferrous (iron) parts and is popular in the aerospace industry.

It's a lower-temperature process that results in a hard case, which itself is under a compressive load. This has the effect of significantly increasing fatigue resistance as well as delivering about a 5-percent increase in tensile strength. Wear on crank journals is markedly reduced and fatigue life considerably increased.

Another process, apparently exclusive to Hinterlighter in the Los Angeles area, is called "cold casing" and achieves similar results.

One process, now losing favor for environmental reasons is "tuftriding." A few companies that have been able to

So that cranks live, GM goes to almost extreme lengths to ensure the stock damper works. That's a point to not overlook!

This, and the other dampers in the GM replacement range from Professional Products have proved functional and cost effective.

afford the system-clean-up modifications are still doing this. The phone book should steer you to a convenient shop.

The last process is relatively common: shot peening. It must be done right to make the most of it. Find a company that does aircraft cranks and ask for your crank to be shot peened to MIL-S-13165C or any superseding spec.

Crank Vibration Dampers

First let me make it clear that the often-used term "harmonic balancer" is totally wrong. This item does not balance harmonics, nor can anything else for that matter. It is a "crank vibration damper" and, as the name would suggest, damps crank vibrations.

Crank dampers often are viewed as an inconvenient and weighty hindrance to performance when in reality the reverse is true. Having spent many hours damper/crank vibration power testing, I am able to pass on some little-appreciated speed tips concerning dampers.

To see where we're going with this, you need to understand that a modern, computer-based cam profile is designed on the premise that the crank rotates in a smooth and uniform manner.

Vibrations transmitted from the crank to the cam can have a significant adverse effect on valvetrain dynamics. By installing a damper, the back and forth oscillation that the crank nose experiences is reduced and power is increased. Acceleration tests on a nominally 400-hp engine on my dyno were set up to simulate a 3-speed automatic car making a quarter-mile pass. These tests showed that, for an engine turning up to 6,000 rpm, the best stock damper was the largest and heaviest available. This heavy damper not only damped best but also produced lower ETs than a super-light aluminum hub that provided no damping. As RPM goes up from here, the picture changes and, because the energy involved equals $1/2\ MV^2$, the optimal damper gets smaller.

To achieve the desired results, a damper must be in good condition. This means the rubber between the outer damping ring and the damper hub must be near perfect. A functionally used damper is normally inexpensive at the wrecking yard. If you have to buy new, you'll find a stock factory one is pricey. Also, if you intend to race your car, be aware that many sanctioning bodies require the damper be made of steel as opposed to cast iron and that it has SFI certification. This means buying an aftermarket damper. Since the first edition of this book the whole damper industry, in the budget to mid-price or "sportsman"

Favored by pro engine builders, the BHJ damper, though hardly budget, is nonetheless a good value and should be considered for any serious build.

end of the market, has changed. To an extent this has simplified my job in terms of making recommendations. Here is how the things currently play out. For my outright race engines I use ATI or BHJ. The ATI damper is a purpose-built item that is definitely for an up market race application. This is what I have on my Cup Car motor. It does a great job and goes back to ATI every thousand miles for a total rebuild. The BHJ damper, though less fancy looking, is also a purpose-built damper and a little less money. I recommend it for the more costly of the budget motors we are dealing with here.

If I figure the price of all the parts is going to be in the $6,000 or so range and peak RPM up toward 8,000, then I usually opt for the BHJ unit. These are custom built for the job, taking into account all the parts used in your engine. It's a quality damper, and you will pay the price for it. That comment might sound like I have just priced it out of your reach, but sales of the BHJ unit are on the rise and prices are on the way down; so check first to see if you can afford one. Last on the list are the dampers from Professional Products. These dampers are typically priced way less than stock and come with degree markings on them. For applications up to about 500 hp, they are approved by Scat for use with their cranks.

Now take note of this. A crank manufacturer is not going to recommend a poor damper as it impacts the reliability of their product. So the Scat approval here makes these dampers worth serious consideration. At the time of this writing I have had some five years' experience with these Professional Products dampers and have had no problems. Dyno results look like they work at least as well as a more expensive stock damper. In this context we have to say that makes them at the very least, a super cost-effective replacement for the stock damper.

If it sounds like I am evading the issue of function or performance here, then that's not the intent. The problem with elastomer dampers is that they are tuned to damp at a certain frequency. The stock dampers are tuned to damp whatever crank, rods, piston, clutch, and flywheel or torque converter combination the factory may have used in the original build. For an aftermarket damper, the manufacturer has to deal with the possibility of the end user having a range of parts. The torsional stiffness, and consequently the frequency of the vibrations, of an aftermarket heat-treated forged crank can be quite a bit different from a cast crank. What this means is that your selection of damper size is not quite as simple as just looking through a catalog and arbitrar-

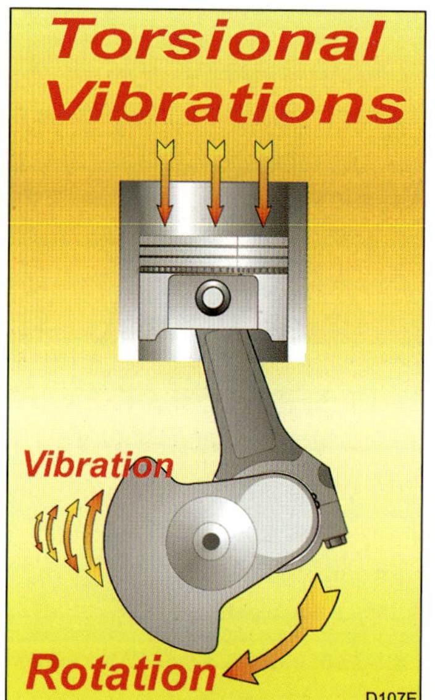

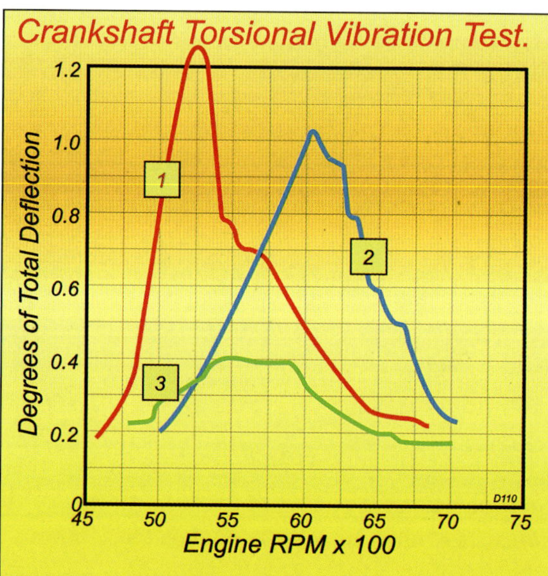

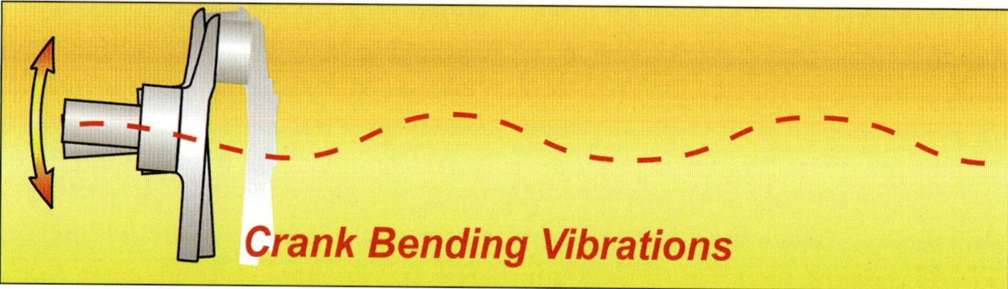

The drawing at top left shows exactly what torsional vibrations are. These are not the only periodic vibrations a crank undergoes. There are bending loads as depicted in the lower drawing. These bending loads become measurably greater when the crank is of the external balance style. An elastomer damper provides a measure of damping on these vibrations as well as torsionals. The graph, from results of my dyno, shows just how much the crank can flex without the benefits of a damper (red curve). On a 350-inch engine the flex, at the journals, amounts to well over a 1/32 inch of back and forth motion. This is like applying about 600 to 700 ft-lbs of torque to the end of the crank and reversing it over 200 times a second! The blue curve on the top right graph is for a mismatched stock damper. The green curve shows what a really effective damper can do.

ily choosing a damper. My recommendation here is to buy the damper with the crank and preferably the rotating assembly. This is a service Scat offers and who should know better what dampers will work with a combination than a company that deals with thousands of such a month. Failing that, call Professional Products and buy the one they recommend.

Do not overlook the fact that a damper must be SFI approved if it's intended use is for a race engine. For the street Professional Products regular dampers are fine, but for race use you will need to get an SFI-certified damper.

Before parting company with the damper subject, I feel I need to better quantify the difference in power an effective damper can make. If you figure on a nominally 500-hp engine as a starting point, then going from a zero damper to having 85 percent of the torsionals damped out, the power difference is typically 12 to 15 hp as measured under accelerating conditions. Any damper that is even half-way effective

Compare this early-style forged rod with the powder metallurgy rod shown below and you will see how archaic it looks in comparison.

The stock powder metallurgy rod seen here has been pin honed to fully float it. When paired with the KB cast piston, it is about as cheap as a good setup gets and will hold up at 480 hp or so.

will allow the engine to outperform a lightweight hub that has zero damping. With that I hope I have convinced you that a damper is a component that should never be overlooked.

Connecting Rods

There are many things I could say about connecting rods, but there isn't space in this book. As a result, I'm telling you only what is needed here.

First let us consider the stock rod. When I wrote the first edition of this book, I went into great detail on the selection and reconditioning of stock rods. The advent of the stock replacement rod has made the use of reconditioned rods almost pointless. These days it costs almost as much to recondition a set of rods as it does to buy new and sig-

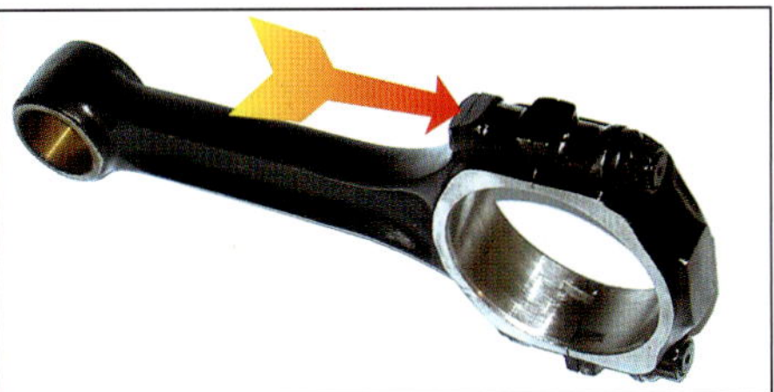

Here is a thru-bolt I-beam rod that I reworked for a 383 back in 1999. To get it to clear the cam, it had to have the bolt shoulder ground some. This has subsequently started to show signs of distress when checked out at about 75,000 miles.

nificantly better rods. This does not mean that you absolutely have to buy after-market rods to build a hopped up motor but it is close to that.

As far as the stock rods go there are two rod groups from which to make a choice. First there are the old-style forged steel rods that were used up until about 1987.

From there on the rods were a much more high-tech deal. Instead of the regular forging technique, GM adopted a process first pioneered by Porsche. The technique is known as Powder Metallurgy Forging (PMF). This process involves fine powder steel alloy being pressed into an accurately finished form of a rod but at a slight amount larger than the finished size. The pressed powder rod is then heated up to a high forging temperature, placed in a net-sized die, and stamped with a great deal of force. This causes all the particles to mash together to form a solid high-grade steel-alloy finished part. These rods look good and in practice they prove to be significantly better than the old forged steel rods. These rods are far better than any of the previous rods and do make a good choice for a rebuild up to about the 480-hp mark if you do not have the budget for a new set.

All the forgoing on powder metallurgy rods is to steer you away, as far as possible, from the early forging. If you have virtually no money then those early steel forged rods will work. Years of use when nothing

else was available have proved that, but be aware that they finally end up being the Achilles heel of any builder capable of finding real horsepower, and after reading this book, that could be you!

Even when prepped out on a cost-no-object basis, we still broke stock rods but at higher levels than stock. An engine with a broken rod is not a pretty sight, and it is not cheap to fix!

If you are forced to use stock rods of the older steel forged type, then you can do as I have done in the past. Here there are three scenarios: First, you may have access to a bin full of used rods at your local engine reconditioner. This is the best situation to have, as there is such a big variation in the rods. Second, you may be using what you have from an engine you already own and have stripped. Third, you may be buying all of your parts from a discount house.

If it's the first instance, where you have a bin of rods to choose from, here is what you need to do to get eight decent stock rods. First select a set of eight rods and maybe a spare based on the following parameters:

1. There are no rusty rod cores.
2. Choose only those rods with the smallest balance pad as they are typically stronger.
3. Only use cores that have a centrally placed pin bushing within the forging

At this point you will have some decent forgings. If the rod journal bores check out OK and the bolts are in perfect order, you have a set of the most basic rods for a hopped up motor. The minimum op here, if full floating pistons are to be used, is to hone the pin bore to a 0.001-inch clearance fit on the wrist pin.

You now have a set of rods but to make them any good for even a half-way serious build they will need to be reconditioned and have ARP bolts installed. By the time you have paid out for this you will have about 75 to 80 percent of the cost of Scat's least expensive rod. At this point you should consider this: the Scat rod is made from 4340 steel and delivers a 50-percent increase in strength over the stock rod. It also comes stock with ARP rod bolts, is bushed at the pin end, and can be had at 6-inches long for a better rod/stroke ratio. You get all this with a rod that is still about stock weight, and for only about 25-percent-more outlay than a totally reworked stock-length forged rod.

Another option you might want to consider here is the stock powder metallurgy rods. These are sometimes on sale from one or other of the big parts houses at about the same price as a total overhaul of an earlier forged rod. If you are really cramped for cash and $30 to $40 makes a difference, these rods are a good option. Remember, however, they are a press fit at the pin end so you still may end up spending another $30 on having them honed to a full floating spec, which ultimately depends on the style of piston you end up using.

If you are staying with a stock-style piston, which has no means of pin retention, then the press fit pin has to be retained. However, most aftermarket pistons have circlip pin retention built into them so they can be used with full floating pins or press fit pins. It used to be the case that to convert to a fully floating pin end on the rod they had to be bored out to take a thin-wall bronze bushing. That is not the case any more as modern oils have made the bronze bushing a redundant element in this application. If you choose to go to the full float-style pin, have the rods honed to give a clearance of 0.0007 to 0.001 inch.

Let's sum up where we are at this point. My feelings here are that, except for a regular re-build, it is just not worth even using those early rods. Why? Because at the end of the day you still have an early rod that was poorly made by current standards and can, even with the cheap speed equipment we have today, be overloaded and consequently unreliable. If we are considering the pow-

der metallurgy rods, a whole different scenario applies. If you end up with a set of these rods in obviously good condition, then by all means use them within the limits of their capability. The best guidelines I can offer here shake down like this: If the rods are in perfect condition, have the pin end honed to convert to fully floating. If the spec you are building looks like it will be more than 480 hp and 6,750 rpm, then get used to the idea you will need to go with aftermarket rods.

Lightening and Rod Balancing on the Cheap

In the first edition of this book I went into great detail on lightening and balancing stock rods. I showed how it was possible to take a stock 610-gram rod and reduce its weight down to 560 grams and still have a stronger-than-stock rod. This all took a great deal of work and effort. With these early rods the forgings were all over the place and putting together a set of the best possible rods (based on stock forgings) was nearer a career than part of the job of building an engine. These days the only stock rods I use are the later, powder metallurgy rods. These are so consistent that, other than honing out the pin end, no work on them is required. Their weight from one rod to another is such that you don't need to worry unduly about balance—it's good enough as is. That cannot be said of the earlier rods. Because the PMF rods are so much better and the price of entry-level aftermarket rods has come down, re-working the early rods is a waste of time. But before finishing any discussion of stock rods, let's consider what makes a rod break because it has a distinct bearing on the power levels a stock rod can deal with.

Even at the RPM we're likely to turn a modest-budget engine, the inertial loads account for more rod-breaking stresses than gas loads. This means, as far as possible, making sure there's no excess reciprocating component mass along for a free ride. The rule here is to keep piston,

pin, and rod weight to a minimum consistent with the RPM involved. The PMF rod does not lend itself to lightening. The factory was able to make these to dimensions that are sufficiently close to what was needed. This means the PMF rod has metal where it is needed so there is little room for any metal removal without incurring a weakness. Bearing in mind that it is RPM rather than just HP that breaks the rods, we find that if power is gained by virtue of nitrous injection, then the rods can be used to significantly higher HP figures. I have used the stock PMF rods to about 575 hp in half a dozen budget nitrous engines and have not broken one in the period from about 1995 to the time of this writing (2008).

Rod Upgrades

In terms of upgrading to a stronger and possibly heavier rod: though safer, it's as bad to over-rod a performance engine as it is to under-rod it. The target is the lightest rod we reliably can get away with.

The first Chevy aftermarket rods were built with the pro racer in mind and were nothing short of expensive and as heavy as required for life in a high-output, endurance-race motor. Their added weight over stock also

Although slightly more money than a basic through-bolt rod, these cap bolt Scat rods are only a little more money and can be cut for cam clearance at the shoulder with no ill effects.

called for the best in cranks. At 700 to 750 grams, such rods are heavy enough to warrant, in most cases, expensive heavy-metal balance jobs. Sheer demand and quantity production since the late 1980s has brought about the production of lighter rods for applications less stringent than all-out race.

These rods fall into the "sportsman" category. They are usually made of the same 4340 material that steel rods are produced from for racing, but are lighter forgings and not machined all over. This is good for the budget-constrained engine builder because: A) we do not want to put any more reciprocating mass into the engine than necessary as it will stress our crank more; and, B) they save on heavy-metal-balance jobs, costing about half the price of a set of race rods.

Rods: The Final Choice

I have used rods from a number of companies over the last twelve years. Experience with such has brought me to a convenient conclusion: namely, that Scat produces among the best bang for the buck for regular and stroker applications for non-nitrous power levels up to about 625 hp and 800 hp with nitrous. Since about 2004, that, with exception of a few K1 Technologies (another company I can recommend) engine builds, is all I use.

At this point I am going to make the assumption you will be using a Scat rod. The question is, now which one is right for the application at hand? The cheapest Scat rod, the 4340 standard I-beam thru-bolt is available in both 5.7- and 6-inch length with bushed or press-fit pin. This rod uses the same bolt style as a stock rod, which is commonly referred to as a thru-bolt. This rod is good in a 350, but for a stroker application it needs the shoulder, and part of the shoulder end of the rod bolt grinding to clear the cam. Although this can be done I have seen evidence that over the long haul the rod bolt head starts to fail.

The bottom line is that in a 350 this rod is a great buy and will serve well, but

Currently my number-1 recommendation for a serious but budget-constrained small-block build—this Scat Premium 7/16 rod is light, strong and has the minimum of bulk around the crank journal end. This means it provides more clearance to the block and cam when used in a stroker application. With most blocks, but especially the later post-1987 blocks, this rod will usually clear cam and block when a 3¾ stroke is used. This makes it a drop-in deal for a 383 build.

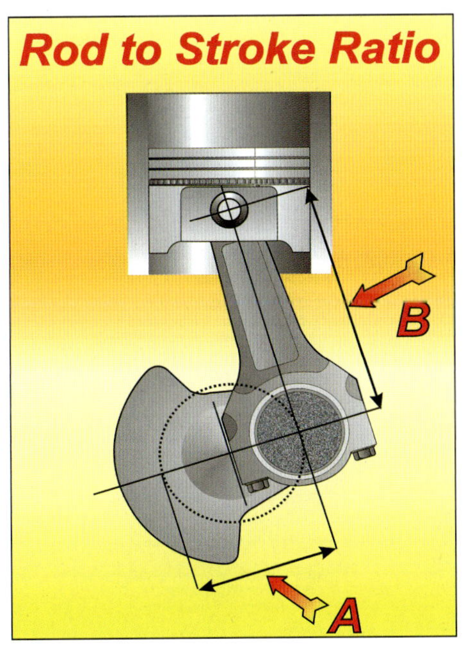

The definition of the engine's rod/ stroke ratio is the center-to-center rod length (B) divided by the stroke (A).

for a stroker motor it is best to step up to, at least, Scat's next offering. This is its 4340 Premium cap bolt rod. The "premium" moniker makes this sound like an expensive rod, but in actuality it's only about $20 or so more than the basic rod. What you get here is a cap bolt (ARP)

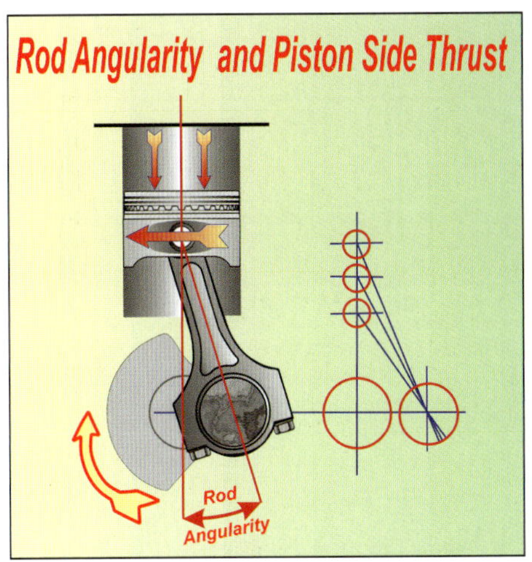

The greater the rod length, the less angularity it goes through during the rotation of the crank. Less angularity is better.

instead of a thru-bolt-style fastener. This means the rod shoulder can be ground the small amount required to clear the cam when a 3.75-stroker crank is used. This rod is also available press-fit or bushed at the pin end and in 5.7- or 6-inch center-to-center lengths. This rod is a really good choice for an entry-level 383.

Last on the list is my favorite. This is the Premium 7/16 rod. It is a totally different forging to the previously mentioned rods and comes with ARP 7/16 bolts. Although it's barely any heavier than a stock rod, it is the strongest of the three rods under discussion. The best part of the deal, though, is that this rod is profiled at the shoulder to clear a stock-base circle cam in a stroker application.

Although it is about 50-percent-more money than Scat's cheapest rod, it is the one to strongly consider if you intend to build a serious high output motor with either a stock or stroker crank. In addition to the normal 5.7- and 6-inch lengths, it is also available in 6.125- and 6.2-inch lengths.

Length: What to Use

The rod lengths available to you range from the short 5.56-inch center-to-center length of the stock 400 rod to 6.2

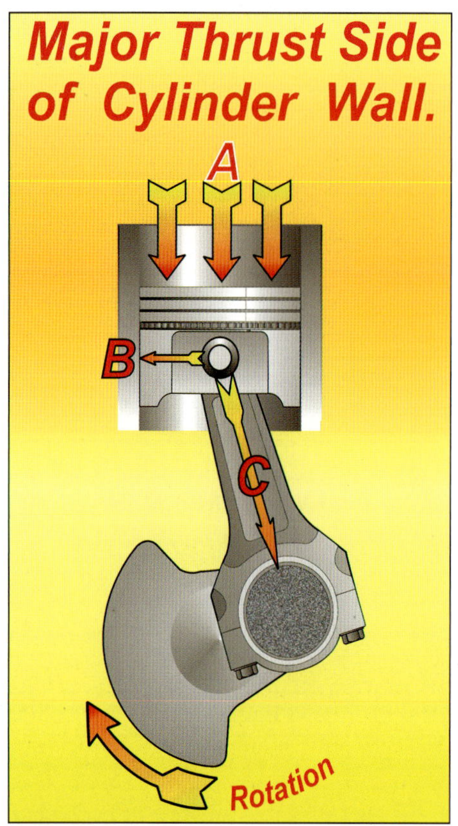

The force produced by the gas pressure (A) splits into two components. These are the force that turns the crank (C) and the component into the cylinder wall (B). The higher B is, the greater the friction between the piston and the cylinder wall.

inches of the Scat Premium 7/16-inch rod. So what should you go for?

In essence, this all hinges on achieving the best rod/stroke ratio the definition of which is shown above. Our main goal here is to minimize piston-to-skirt friction when the rod is at its greatest angle to the bore.

In practice this means utilizing as long a rod as possible within the confines of the engines we are dealing with here. A rod/stroke ratio of about 1.8 to 2:1 would be about optimum but with the block heights and stroke lengths we are using here, that is not a practicality. What this means for the most part is making the most of what will go in, and for the most part this will be a 6-inch center-to-center rod.

If you become involved with engine modifications to any real extent, sooner or later you will hear the comment that short rods make more low speed torque. The reason behind this is due to the fact that the shorter the rod is, the longer it takes to move any given distance either down or up the bore around bottom dead center. This in turn means that the piston is less distance up the bore when the intake valve closes, so it traps more charge above the piston and thus delivers more torque. This may be so, but offsetting this is the fact that the rod's greater angularity pushes the piston into the bore with greater force on the power stroke. This in turn increases the piston-to-cylinder-wall friction. So it looks to be a trade-off here.

You will get plenty of engine builders to tell you that, in their opinion, the short rod is better for an engine that may never see the topside of 5,500 rpm. I am not in the business of opinions so I dyno tested a 383 with a 5.56-, 5.7-, and a 6-inch rod. This took three different rod and piston assemblies to do so and was not a cheap test by any means. The bottom line is that the 5.56 rod produced a lesser output right down to 2,200 rpm, which was as low as I could test. The 5.7 rod was better and the 6-inch rod the best. The differences were not huge by any means but what did show up was worth the effort to make the change. In round numbers, the 6-inch rod was about 7-hp up on the 5.56 and the 5.7 was about midway between the two. One aspect of importance that would not show on a power curve was the fact that the longer the rod, the mechanically quieter the engine ran. For that reason alone it's worth using only a 6-inch rod in any build.

I have talked as though a 6-inch rod is "it" for all applications. Sometimes that is all that can be squeezed in. An instance here is a 4-inch stroker conversion into either a 350 or 400 block. About the shortest practical compression height for a piston is 1 inch. This means that with a 9.000 block height the longest rod that can be used with a 4-inch stroke is 6 inches. If you are building a 3.75 stroker then, a 6.125 Scat Premium 7/16 rod can be used with a piston having a 1-inch compression height.

The variety of budget high-performance pistons available to the hot rodder has never been better. Shown here is just a sample of KB's cast hypereutectic offerings.

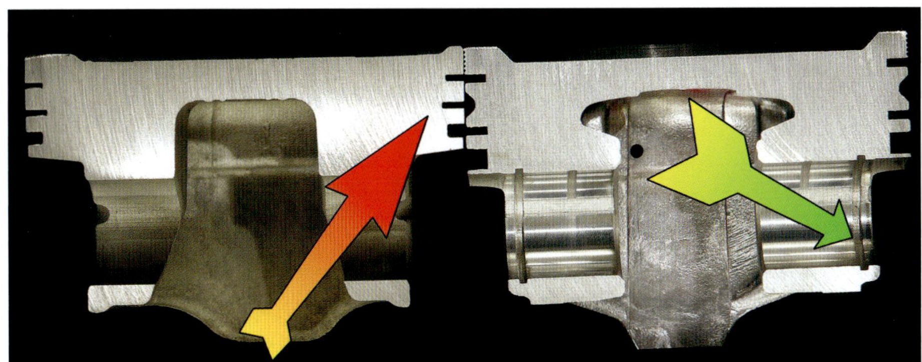

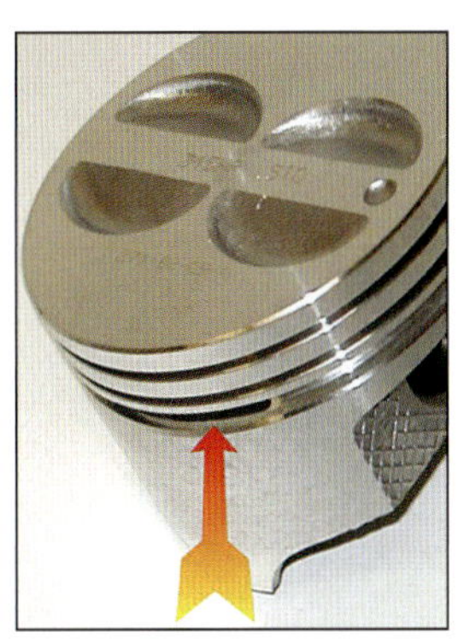

By adopting a slot in the oil control groove as shown here, heat transfer from the piston crown is considerably reduced. This cuts skirt expansion and thus allows a closer fit to the bores. The price for this is a piston that can break through the oil ring groove.

A typical stock post-1987 forged piston is on the right and a cast high-performance KB on the left. Check the area for comparison indicated by the red arrow. What you see here are rings on the OEM piston that are smaller and narrower than a typical low-buck aftermarket performance piston. Although these narrow rings are more expensive by far than the 5/64 wide compression rings (which are really low cost) used on the piston on the left, the fact that you already have a good piston means only buying a set of medium-priced rings instead of a low-cost set of rings and a set of pistons. If you choose to stay with the factory forged piston, you will save money and have a piston good for about 500 hp. There are, however, a few downsides to retaining the stock piston. Because the factory utilizes a press fit pin, you are kind of stuck with a 5.7-inch press fit pin rod. The KB piston on the left can be used either press fit for the pin or full floating as it is equipped with clip grooves (green arrow) to retain the wrist pin.

Oil return via drilled holes in the oil groove rather than a heat dam slot, provides for a much stronger piston. This is the design to go for when selecting a piston for high-performance use.

Pistons: Type to Use

There are two choices of piston types for an engine rebuild: forged, or cast. For a performance engine, the most popular is the forged piston but that is slowly changing. The basic differences between these two types are related to both function and economies of cost. A forged piston is tougher and often has higher strength at temperature. A cast piston, which usually has a high or very high silicone content, may, depending on the alloy, be as strong but is harder and, as a result, experiences less skirt and ring groove wear. Also a cast piston expands less than a forged one so it can be run at closer clearances for quieter operation. Lastly, in most instances, they are cheaper to make. I intend to walk you through the aftermarket piston maze but before going there it is well worth looking at what your engine may have been equipped with from the factory.

The push to get ever better mileage has lead to many upgrades on the pistons that GM installs at the factory. Discarding these with no thought as to what benefits they may have is not a smart or cost-effective thing to do, unless you know for sure the engine will need new pistons.

In a bid to cut internal engine friction, develop a better gas seal, and last longer, GM revised a lot of the pistons used on post-1987 engines. The pistons you have in your motor should be checked out for the features shown above. My recommendation is that if you are on a really tight budget, then it is OK to stick with the stock factory high-performance piston, the stock crank, and powder metallurgy rods.

Pistons: Aftermarket

When choosing a piston style that fits your budget, you could be confronted with many different piston designs. What is ultimately chosen will depend on the power level being targeted. The requirements of a stock piston are that it is cheap and, at the power level it is intended to run, is reliable. Also, it must operate without any audible piston slap at any temperature from –30 degrees F to as hot as the engine gets during full throttle operation. Such pistons commonly have a slotted oil-control ring groove. Other than to act as a route for oil return from the oil control ring back to the pan, the purpose of this slot is to stop some of the heat that would otherwise expand the skirt from actually getting there. This allows a closer fit to the bore for a quieter

KB's low-cost Claimer pistons can by identified by the "claim" logo on the pin boss strut as shown here.

Here is a KB hypereutectic piston that has delivered positive results in a 350 when paired with a 50-cc chamber head for a 10.3:1 CR.

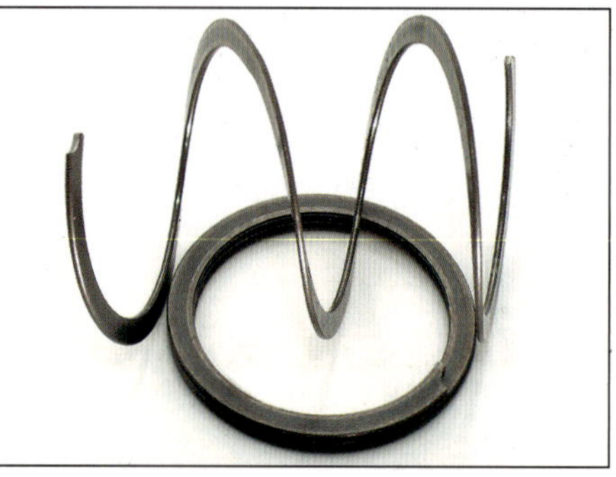

The Spiroloc pin retainer presents the biggest challenge to the novice engine builder. The key to easier installation is to stretch it out as shown here, then wind it into the groove.

Before installing a piston, check that the snap rings are fully home in their grooves and that the oil rail expander ends are correctly butting and not overlapping. Fail here and you will pay the price!

The arrows on this cast piston indicate the renforcing struts making these pistons stronger than the alloys used might suggest.

operation. As good as this is for quiet operation, it can be the death knell for a performance engine if a piston of this style is used, unless such a piston has increased cross sections elsewhere to compensate. Some forged aftermarket pistons have a slotted heat-dam oil groove, but these pistons have a thicker cross-section joining the pin boss to the skirt in order to compensate.

If you are using a cast production-type piston of this style because of cost constraints, then be aware that 5,500 rpm and about 375 hp is typically the limit for these pistons. If the pistons are overused, the top of the piston will part company from the skirt section, rendering it useless other than to give you more practice at fixing blown up engines.

For a performance piston, the usual means to return oil to the pan is via oil holes in the bottom of the oil groove. This does mean that more heat gets to the piston skirt, causing it to expand more. That in turn will call for an increase in skirt clearance and probably a little noise on start up. But that is a small price to pay for a much stronger piston.

Over the years I have used most brands of pistons, but in the last ten years or so I have settled on a few brands that meet my needs on a range of engines from really cheap to mid-price but still pretty serious race stuff. These brands are Speed Pro, KB, Ross, JE, Mahle, and Wiseco. The only Speed Pro I use are the cast hypereutectic performance-oriented pistons.

The KBs I use are split 50-50 between cast and forged, while for companies other than Speed Pro, it's all forged stuff. Let's start with the cast hypereutectic pistons first.

A question that you may well ask here is, "What does hypereutectic mean?" Essentially, "hypereutectic" means an alloy that contains more of an alloying element than will dissolve in the parent metal. The principle alloying element for an aluminum alloy intended for piston usage is silicon, and for that you could read "glass." Up to about 12-percent silicon will dissolve in aluminum, and any excess above that disperses "as is" throughout the material in the form of crystals. The silicon increases the strength of the aluminum and also makes it considerably harder and to some extent more brittle. Although a hypereutectic alloy lacks some of the toughness of a forged alloy, it has more than one redeeming factor that allows its use to higher power levels than might normally

Study the underside of this forged KB piston and you will see how the form is tapered outward toward the open end so that the forging punch can be removed.

Here is a flat-top KB forged piston. It's relatively light, strong, and cost effective. I use these in about one out of three builds.

Many of the older-style pistons use a longer pin (left). Newer designs with closer pin bosses use the shorter, stiffer pin (right), which saves a lot of weight.

be expected. The first of these is that as a casting it is easier to put metal exactly where it is needed to hold out against the loads applied during operation. This means thicker cross sections at key points can be employed. The bottom line is: so long as it is not overloaded, this type of piston makes a good choice for many builds up to about 500 hp. I actually got to do some testing on the Speed Pro pistons prior to their introduction. I used them at 510 hp in a naturally aspirated engine for about a month on the dyno, and a nitrous engine with about 50 nitrous pulls on it at as much as 560 hp. Nothing broke.

As for KB's pistons this company has really embraced the task of producing a near bulletproof hypereutectic piston at a super low budget. KB's Claimer race pistons may not be the prettiest you will ever see, but they are tough players and among the cheapest out there.

KB forged pistons also figure high on my list of cost-effective pistons. Although they are more money than the cast variety, they do allow a significantly greater power loading to be employed without the fear of failure. If we are talking in terms of nitrous, and by that I mean nitrous done right, then 850 to 900 hp is fine. One point you should check though, regardless of the piston used, is the pin clearance within the piston—this

is especially important if the application is still using a press fit pin. To avoid a pin seizure when the nitrous is used before the combustion heat has brought the pin area up to full operating temperature (often the case at the drag strip), the piston pin bores need to be honed to a minimum of 0.001-inch clearance on the pin. Also when choosing a piston, especially for a stroker motor application, be sure to get the lightest your budget will allow. Ignore this and you could be paying much more for a balance job.

At a little more money, the Ross, and more recently, the German Mahle

The Mahle piston is supplied with reduced section lightweight rings and is truly an up-to-the-minute design.

Piston coatings used to be Pro Stock stuff but reduced costs and do-it-yourself suppliers have moved it down to the everyday hot rod build.

Here is a Wiseco piston for one of my high-end budget motors (that is about six grand as of 2009). It has a ceramic-coated crown, anti-friction skirt, and 0.043 wide compression rings.

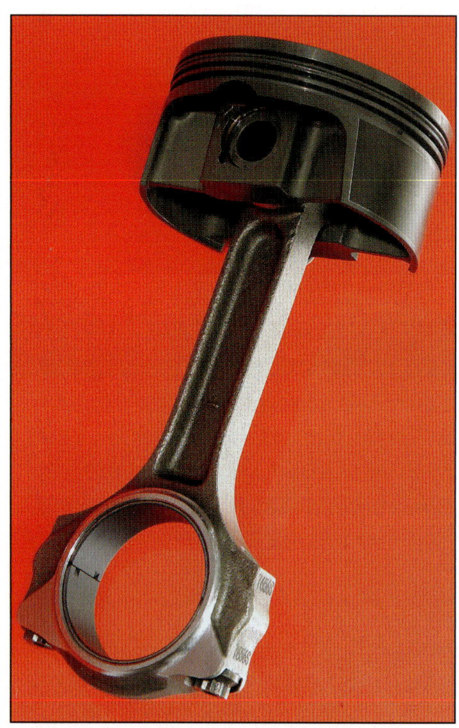

What you see here is a highly functional rod and piston assembly for a 600-hp build. The Mahle piston, Scat rod, and Calico coated bearings are all off-the-shelf parts.

(Mah–lee) pistons figure strongly when I am watching weight with the intent of achieving internal balance on a stroker motor. There is not a lot I can say about the Ross pistons that I have not said before. Two decades of use in a couple of dozen street/strip engines up to 800+ hp and never a problem.

The Mahle pistons are a fairly new deal and I have experience of only a half dozen or so builds over a two-year period at the time of this writing. That said, I am impressed with the results and the degree of high tech delivered for the price paid. These pistons have a lighter ring package and come with an anti-friction coating on a strut-supported skirt. In short: they are tough, stiff, and light and the efforts at cutting friction during operation appear to have paid off.

For my higher-end pistons, and we are still talking only about $5,500 per motor for total parts expenditure, I use JE and Wiseco. At the time of this writing I am favoring Wiseco here because of the convenience of skirt coatings as a standard with the option of a metallic/ceramic thermal-barrier coating for the crowns. The coating is a type I have had good results with in the past, especially with nitrous motors. In fact, since using this coating (more than 10 years as of 2009) I have never come close to damaging a piston from thermal abuse (i.e., lots of nitrous). Another factor I like about the Wiseco piston is that the turnaround time is good and they are not that much more money for a custom piston. Buying a custom piston allows me to specify what I want for rings and this can prove a worthwhile power asset in itself.

Piston Selection

Before you choose a piston, read Chapter 6. The reason for this is that the piston constitutes half the combustion chamber form. It is, for a novice, all too easy to select a piston/cylinder head combination that is far less than optimal. Here are a few ground rules: First, choose the lightest piston possible consistent with budget and usage. Also, choose a piston requiring as little piston-to-cylinder-wall clearance as possible consistent with use and piston type. Note here that hypereutectic pistons usually need less clearance than forged, so, in the majority of cases, they will run quieter. After you have studied Chapter 6 select a piston that, in terms of CR, engine displacement, and cylinder head chamber volume, allows for a flat- or reverse-dome piston crown. If you must use a raised crown piston to get the CR up to where it's needed, then limit the dome height to no more than 0.150 inch. If you intend to use full floating rods, be sure to get pistons with the appropriate method of pin retention; i.e., not press fit. In order of ease of fitting: the double lock snap rings are the easiest, followed by the round wire clip (Mahle), then the Spiraloc (which can be a pain until you get the knack of installing).

Size 5/64 rings are used on a lot of the cheaper pistons. This is a size that was common on stock pistons up until maybe the early 1970s. This size of ring is ridiculously cheap at about half of the cost of the cheapest 1/16-inch-wide ring. These wider rings are OK for regular use up to about 450+ hp, plus they work OK with nitrous. The down side is that without proper care they do not last as long—but there is a fix, as we shall soon see.

Rings

For those on a really tight budget, you can source "no label" 5/64-inch-wide rings intended for the high turnover rebuilding shops. These cheap 5/64 rings will work well for an engine not expected to turn over 6,000 rpm with a 6,500 max. If the build is a stroker motor, drop the rev limits for wide rings by 500 rpm. These wide rings are almost always made of a relatively soft ductile iron and will bed in quickly. They will also wear much more quickly unless suitable steps are taken in the upper cylinder lubrication department (see nearby Sidebar, "Minimizing Engine Wear"). A 1/16- to 3/16-inch ring package is better, especially if the top ring of the set is a moly plasma ring, but these will cost more.

If you are building a stroker motor, then the reduced width of the ring belt available for the rings almost dictates the use of 1/16 wide compression rings. If the budget is sufficient then it is a good idea to consider a Total Seal ring in the top groove. Since the first edition of this book I have done a lot of back-to-back testing with this style of ring and believe me, they are about the best on the general market. If there is anything better, the F1 guys are keeping it under their hats! There is not room here to detail all the tests I did over about a four-year period, but you can get a complete report on my website at MotorTecmagazine.net.

Minimizing Engine Wear

Wear is the performance engine's enemy, especially when it affects the cylinder's ability to produce a gas-tight seal. My work involves dyno testing with relatively costly engines that must maintain a consistent baseline over periods as long as three months. If the valve seal can be maintained and ring and bore wear countered, the engine will produce like new. I have dyno tested many "wonder additives" and few are worth the money. As a bore-wear inhibitor and fuel system/injector cleaner, ACES is an exception.

Using two engines, 250-hour wear tests were conducted on my dyno. The first engine used only a non-lead control fuel direct from the refinery. The second engine used the same fuel with ACES added. After each engine was run 250 hours it was stripped and various parts measured. Taking the least worn cylinder of the engine running on untreated fuel and comparing it with the worst of the ACES treated engine revealed that this additive reduced bore wear by a minimum of 600 percent. Also, it was found that exhaust-seat recession on the non-hardened seats of the test engines was about halved.

Subsequent tests on road and race engines have shown that if ACES is used at about 25-percent higher concentrations at break in, and as little as half the recommended dose thereafter, bore wear over long periods (300,000+ miles) is close to zero. The greatest advantage is the virtual elimination of the formation of a ridge at the top of the bore where the top ring parks at TDC. This eliminates re-boring, other than from catastrophic component failure. As for break in with ACES, this appears to take 50- to 100-percent longer. However, it

produces a lower-than-usual-friction bore finish.

ACES produces these results by turning into a high-grade synthetic lubricant during the combustion process. It coats the cylinder wall with a layer of lubrication just a few molecules thick. The result is the rings always ride on some sort of film. Since it is replaced every cycle, the rings never run dry. This action reduces friction and increases power. Half a dozen tests show power gains from reduced friction in the region of 1 to 1.5 percent. That's 4 to 6 hp for a 400-hp motor. For extensive 20+-year test data on this product go to MotorTec magazine.net. To purchase, contact BND Automotive.

Be aware you will not find ACES in regular parts stores. It is not a highly diluted snake oil mix for stores. It is a commercial grade, concentrated additive sold mostly to big fleet users and oil refineries and such companies do not waste their money on stuff that does not work—but neither do I! What you see here is enough to last a typical user 15,000 miles. The dispensing bottle, with its measuring compartment, holds enough for four to five fill-ups. In addition to cutting wear it also helps boost octane, cleans fuel systems, and de-waters fuel.

This drawing shows the basic principle of operation for the Total Seal top ring. High-pressure gases coming through the main ring squeeze the ring assembly onto the bore, leaving no direct passage for leakage.

Balancing

Let's start with the stock rods here. First, if you have the powder metallurgy rods you will not need to balance them because they are, for the budget builder, close enough as is. If you have the earlier rods, then life won't be quite as easy. These are selected as a matched (roughly, that is) group and if you selected rods from a parts bin, figure they could be a ways off. Three ways to deal with this are open to you: First, you can dump the old style rods and find some of the new PM rods. It may cost a little more but at least you also get a decent rod in the bargain. Second, you can pay a shop to balance the rods you have, but be sure they warrant putting that money into them. Third, you can scrap the stock rod deal right there and get a set of Scat's cheapest rods because they are already balanced by selection to within 2 grams. Pistons are also usually close enough to each other that any more-precise balancing is one step away from gilding the lily.

Let's take a look at the crank now: what we need to know here is whether or not the crank is OK with the factory balance. Here is a way to check that out: First, you need to know the weights of the pin end and the rod journal end (big end) of the rods. If you have the rods I recommended, it is written on the box; if not, you will have to go to a shop with a rod-weighing fixture and have a rod weighed. At this point add 50 percent of the weight of the piston assembly and the pin end of the rod to the rotating weight (the rod journal end). Don't forget to include the weight of the bearings and 2 grams for oil. Now double the number—if this works out to be between 1,850 and 1,900 grams, then you are OK because the stock 350 crank is balanced with a bobweight of 1,870 to 1,900 grams. The most important part of balancing is that the piston assemblies are all the same and that each end of the rods are likewise—the same. Last, on the must-do list, is to have the balance factor (over or under) the same at each end of the crank.

The forgoing makes balancing seem like it is an operation without some open-ended options. How so? Optimal balance occurs when the bobweight used is 51 percent instead of the 50 percent normally used to represent the mass hanging on the crank's rod journals. If the rod/piston assembly is lighter than stock, it follows that the counterweights will be too heavy by a small amount. This can be an advantage. In highly stressed engines the crank assembly is deliberately over-balanced so that, as the piston travels down the bore, the downward force of the gas pressure loading the main bearings is countered a little by the upward force of the heavier-than-normal counterweight. About 50 grams of overbalance can be used before there is any detectable loss of smoothness. Even if the difference between your crank's theoretical and stock weight is greater than this, all it will mean is that the engine shakes a little more than we might like. Whatever it does, it is still likely to be less than a balanced four-cylinder engine!

Forgoing balancing does not mean that there will be more destructive loads present. What it does mean is that whatever forces are generated internally at one journal may not be fully countered by those at another. The result can cause the engine to shake due to the difference in these loads. The difference between RPM-to-failure on a balanced engine and one that is not is very minimal.

So should you balance your rotating assembly or not? If the budget is there to do so, then go ahead. Be aware that there is more to be gained from a balance job on an engine that requires an external balance damper and flywheel. But let me remind you that, at the end of the day, the best deal is to get the entire rotating assembly (pistons, rods, damper, etc.) from your crank manufacturer already balanced. That is the best deal and your safest option in terms of smooth running.

You can buy a cheap universal application ring compressor, but it only takes one failed installation to make a cheap compressor expensive. My advice, especially for the novice, is to get a precise fit ring compressor as seen here from Total Seal.

LUBRICATION SYSTEMS

The service life of any engine depends greatly on the quality and properties of the oil used. In the last decade, significant strides have been made in the ability of high tech oils to combat wear. At the turn of the millennia you could expect a street-tuned (about 1 hp per cube), daily driven small-block Chevy fed on a diet of top-notch synthetic oil to run for 250,000 miles and still deliver quality performance. These days, if you have chosen the right parts and use the right oil, a half-million miles is most certainly possible.

There's no secret to achieving this longevity. Many factors influence the situation, but the most important are oil changes at regular intervals. 3,000 miles on even synthetic used to be considered the oil change limit for those of us who loved our custom hot-rodded street engines and could not afford to be re-building them every 50,000 miles (or less). I have a Magnuson-supercharged extended-cab Chevy truck and with what I have invested, it had better run 300,000 miles at least. And it will—and so will your Chevy if you follow the advice given here.

Let's deal with the 3,000-mile oil change interval first. Use a good synthetic and a good oil filter and you can very safely double that and more than likely triple it. Even though the good oil costs

What you see here amounts to the heart and life-blood of an engine. Be sure that whatever you do, it is up to par for the circumstances involved.

more, its extended life means you will be paying less overall—and that is an economy right there. The only really effective way to find out if the oil needs to be changed is to have an oil analysis done.

However, that costs about $20 to $25. You could have spent that money on new oil, so is an oil analysis worth even the minimal effort involved? Yes, but not just to establish whether or not an oil change is needed. I have an oil analysis done on my vehicles about every 30,000 miles and this tells me whether or not everything inside the engine is still as it should be. If not, the analysis gives me an early warning.

The next question will probably be: "Where can I get an oil analysis done?"

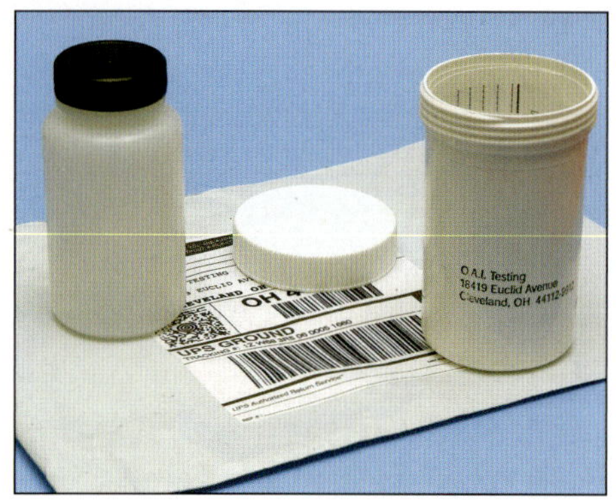

You can get an oil sample bottle like this from any Amsoil dealer. Just fill it and send—postage is pre-paid.

Tracking down a lab that does this can be a pain, but a source that is easy to find and is cheap is through an Amsoil dealer. To find the nearest to you go to Amsoil.com.

Losing ZDP

Since the first edition of this book, there has been a big panic among the cam companies over the loss of Zinc Dithio Phosphate (ZDP) in API certified oils. This additive was removed from the oil because of possible contamination of catalytic converters. The ZDP is a high-pressure lube that is close to essential for the survival of a flat tappet cam and lifters—especially at break-in.

Without ZDP you would have to run a cam with very mild acceleration rates for it to survive. A cam intended to be used with ZDP will last only a few hours without it. The good news is that although those oils at your local parts store may not have ZDP, you don't have to buy oil that meets API requirements. Most of the companies that brew race-orientated street oils are not concerned about getting API certification. They are more concerned about power and having your engine last because their reputation hangs on it. If your engine is going to utilize a flat-tappet cam, and there are certain occasions when it is your best torque/power option regardless of cost,

be sure to utilize an oil with ZDP in it. You will find plenty of recommendations in these pages.

If the right oil is used, the bearing and bores of a small-block Chevy are of little problem in terms of longevity. At the end of the day it really is the cam and lifter contact patch that is the critical lubrication point. As far as oils that will work in respect to an engine equipped with a flat-tappet cam, there are a few I have tried on the dyno, at the track, or on the road. Let's start with a look at what you might use on the track. Here, I did a mega-buck test on the then new (as of 2004) Joe Gibbs Racing (JGR) oil that's used in the team's Cup Car, which at the time was driven by Tony Stewart. Apart from my own tests on the dyno with several different engines and 1,000+ circle track miles, I got to use Gibbs' $1.5 million dyno for a day of testing with one of Tony Stewart's restrictor plate Talladega motors. A comparison was made with some of the best race oils out there and it came out on top, even if it was only by a margin of 1.3 hp. And just in case somebody mentions that you cannot dyno test to that sort of accuracy, then all I have to say is: "You have obviously never used a 1.5-million-dollar dyno!"

But the Gibbs oil might just be a little pricey for many and, though very good at developing a race oil, Gibbs is not the only kid on the block. I have also used Amsoil, BND, Redline, Mobil race oil, Valvoline race oil, and UPM oil. The non-API flat-tappet oils supplied by these companies each have a healthy dose of ZDP, making them ideal for a flat-tappet engine. Also these companies, and especially the Amsoil and BND, have blends that are long life deals on the street and can be used confidently to 9,000 miles.

Power Versus Oil Type

An often-posed lubricant question is whether one oil will produce more horsepower than another or, indeed, whether

or not pouring an additive into the engine will help horsepower.

In most cases concerning additives, it's "buyer beware." Since there are so many variables and products to consider (many of which I have yet to test), I can't be specific in this area, but I can pass on my experiences.

Synthetic-oil manufacturers claim their products produce more horsepower because they are slicker. This may or may not be the case, but evidence from my dyno testing has shown that thinner oil does produce more horsepower. For instance, running a mineral oil at 200 degrees F and then increasing the oil temperature to 220 degrees F will produce at least 3- to 4-hp more in a typical 300-hp motor.

Running mineral oil at 230 degrees F is pushing it near its temperature limits. Certainly, going as high as 250 degrees F is a marginal situation, like treading on thin ice. Synthetic oils, however, are fine to 300+ degrees F. Additionally, their viscosity doesn't change as much as mineral-based oils with rising temperature. This means the oil can be thinner to start with and deliver the benefits in terms of extra power without the need to run such high temperatures.

Within the limits of the dyno's accuracy, my tests have showed no measurable difference between mineral and most (but not all) synthetic oils when both were run at similar temperature-developed viscosities. Many of the test results indicated that, with mineral oils, coolant temperature needed to be 170 degrees F and lubricant temperature around 220 degrees F. It's difficult in practice to get the oil temperature this much hotter than the coolant temperature. When synthetic oils were used, the engine delivered the same kind of horsepower with lower oil temperatures.

A potential power increase does exist when synthetic is used. In an engine of around 400 to 450 hp at 6,000 to 6,500 rpm, a synthetic running 20 degrees cooler looks to be about 2- to 4-hp more.

Read the label—even though they meet API standards a lot of inexpensive oils simply are not suitable for a flat-tappet-cammed small-block Chevy.

This is the number-1 dyno at Joe Gibbs Racing. On the dyno is one of Tony Stewart's Talladega restrictor plate engines. It was in this that I got to test the JGR flat tappet racing oil versus several other top brands of my choice. We used 22 gallons of oil that day!

Such numbers represent the limit of test accuracy on a typical dyno and only can be verified by meticulous tests and averaging a number of test runs.

Using this technique, most synthetics appear to be worth on the order of 0.5- to 1-percent-more power. At the time of this writing, the oils that I use are BND for both street and race, Amsoil for the street, and JGR oil for race applications.

If you are a street or street/strip hot rodder on a very tight budget, these oils may seem like they are a little over the top for your budget engine, but consider this: If you had to struggle to find the cash to build it, isn't it better, as far as possible, to make sure you give your hard-earned investment the best protection possible? Granted, the higher initial expense can be a problem when cash is really short. Fortunately, most synthetics can be diluted as much as 50/50 with a good grade regular mineral oil and still retain better than 75 percent of the synthetics' advantage. For break-in lubrication, a good oil is needed, but a long-life synthetic is overkill because it will only be used for a few hours. For break-in, I use Castrol GTX, Valvoline, and Pennzoil, plus whatever cam break-in additive the cam company calls for. If I'm buying the oil for my vehicles, I use only the best. Internal inspections at 50,000 to 100,000 mile intervals and oil analysis at 20,000 or so miles show it pays off. And for what it's worth: none of my vehicles get an easy life. Even my tow truck gets raced!

This is what I use in the engine of my 740-hp 2002 Cup Car. The guys at JGR hired some of the top oil scientists to come up with this.

Here is an oil I have found to be very effective as proven by oil analysis. Its wear rates are comparable to or better than the top big brand name, along with a slight edge on power.

Whether for the street or track, here are two brands of oil I have found you can depend on for very positive results.

Break-in oils do not need to be expensive or long life—just good for 500 miles. These mineral blends are what I most often use.

For break-in I generally use a System One filter (center), but they are pricey. That being the case, use a Fram or Purolator as seen here.

Oil Temperature

Although lubrication is its primary function, oil is also an important cooling medium. Pistons rely on a significant amount of oil splash cooling on the underside. Although, for windage loss reasons, oil needs to be kept away from the rotating assembly, a certain amount has to be present for cooling purposes.

This leaves us with a trade-off situation. If the oil is too hot, it may be better initially for reduced windage losses, but this can lead to overheating. Even if no oil film breakdown occurs at the bearing, pistons can weaken due to the elevated temperatures and detonation is more likely. Conversely, excessive cooling can lead to viscous drag losses.

Generally, the higher the RPM, the more piston heat there is; therefore, more oil probably is needed to cool the piston. Unfortunately, higher RPM means a greater necessity to keep oil away from the crank because the losses go up with the square of the engine's RPM. Doubling the RPM means four times the windage loss. In other words, if we can cut windage loss by 2 ft-lbs at 4,000 rpm, this will represent about 8 ft-lbs at 8,000 rpm. That may only be 1.5 hp at 4,000 rpm, but this escalates to a whopping 12.2 hp at 8,000 rpm. So what's the best trade-off?

In a naturally aspirated engine, there's usually enough oil splash to adequately cool pistons, especially if synthetic oil is used and temperatures kept to the 190- to 200-degree mark. If temperatures escalate much over that, there may be a penalty of needing thicker piston crowns, slightly lower compression, and so on. If the subject is a nitrous-injected motor, then an entirely different situation exists. Short of paying a machine shop to put in piston oilers, the best plan, if the pistons have not had a thermal barrier coating as described in Chapter 4, is to keep the oil temperature down to the 170 mark. This may call for a bigger pan or an oil cooler. If the nitrous system is only up to about 150 hp and jetted suitably rich, any good oil will get the job done.

Oil Filters

There is hardly any point of using the best in oils if your choice of filter is not equally informed. If you are on the dyno for break-in, be sure to use a good filter for the actual break-in, then use a really good one from there on.

If you are breaking-in on the street and intend to use the break-in oil for 250 to 500 miles before changing it, use a top-notch filter from the get go. As of 2009, I usually use a Purolator Pure One, a Fram filter, or my stainless mesh System One filter for the first half hour on the dyno. When the initial break-in has been completed, the oil is dumped and the synthetic is poured in. The filter is then changed for whatever checks out the best at the time (things can change quickly in this area), and at the time of this writing it looks like Amsoil has the hot ticket in terms of flow and filtration capability. If

After break-in, the Amsoil filter should be about top of your list to use for max engine life.

you cannot locate an Amsoil dealer, an equally good choice should be the aero quality K&N oil filter.

The original 265-ci small-block Chevrolet was introduced without an oil filter. The following year the canister-type oil filter set-up was added and was used until 1968. Subsequent small-blocks used a different adapter in the filter housing and a spin-on filter. That was used until the late 1990s, when production of the small-block as we know it ceased.

For most purposes, the earlier filter was a good deal. In general, the element flow used in early filters was higher, with typically 90 to 95 percent of the oil passing through the bearings also going through the filter.

The later, spin-on cartridge filters eased filter changes. However, the shorter ones have a lower capacity and can't handle the full output of the pump, especially when the oil is thick during a cold-start situation. To accommodate this, an oil bypass valve is located in the filter spin-on housing bolted to the block.

The bypass valve usually lifts off its seat at a maximum of 14 to 15 psi, but when the oil is cold it's not uncommon for the pressure differential across the filter to exceed the bypass-release pressure. This results in unfiltered oil going direct to the bearings. Since the oil pick-up is at the bottom of the pan, where most of the debris settles out after shutdown, we find the filter is bypassing oil at a time when its filtering action is needed most.

At this point you may be tempted to block the existing bypass valve or install a filter adapter with no bypass valve. Before you're tempted to do this, consider the possible problems.

First, plugging the filter adapter valve and using a stock-size (and flow) filter can mean a bigger pressure drop across the filter. This can reduce the pressure fed to the bearings. Also, the pump relief valve is likely to come into operation sooner, so the quantity of oil available at the bearings will likely be further reduced.

In addition, if the oil pump has a high-pressure relief valve, the pressure generated while cold could burst the filter, so now your oil is being pumped onto the street rather than through the bearings. If the filter doesn't burst, consider that the extra resistance to flow through the filter means more stress on the pump and more power to drive it.

What's probably a better alternative, especially if you're working within a budget, is to use the larger small-block Chevy truck filter. With this you can expect to achieve the filtration performance typical of the early canister filter.

If you want to stick with a more or less conventional paper-type filter (as apposed to a steel mesh filter), then Amsoil and K&N Engineering have, as of 2009, among the best on the market. The filtering element of either of these filters flows about 250-percent more than a typical OE filter, while taking out particles up to about 40-percent smaller. The Fram HP filter is also a good, cost-effective choice with about twice the flow of most normal filter units.

If the budget allows, another type of filter I recommend is the System One. This filter uses a stainless-steel mesh and can be stripped for cleaning and reused. The System One is a direct replacement for the stock filter and is available in short and long cartridges. The strong point of this type is the ease with which a debris inspection can be made. An average OE-style filter is good to about 25 to 30 microns, but bypasses so much potentially dirty oil that this can become almost academic. A System One filters to about 45 microns, but because it flows about 800-percent more than an OE type unit, it filters all of the oil.

The reason System One can be a good first-filter choice is that they allow near instant oil analysis after break in, by inspecting what has been stopped by the screen. Since these filters come apart, it's easy to see the debris collected on the filter element. On the other hand, opening up a paper filter is a pain.

The stainless steel mesh of the System One filter allows easy identification of any internal problems after break in.

Attempting to inspect the fuzzy surface of a paper filter is difficult because it camouflages much of the debris. The stainless element doesn't do this. Since test engines are run for a relatively short time in terms of mileage, and are always on a diet of clean, fresh oil, the micro-particle filtration capability of a stainless screen becomes academic, since it would do in most cases on a race car. For street vehicles where oil changes may be as much as 9,000 miles apart rather than at about every 30 to 35 miles, I use Amsoil or K&N paper-element filters.

Oil Pumps

The factory stock small-block Chevy pump is effective and reliable, but not perfect. It's common practice to install a high-pressure, high-volume pump. In many cases this is simply not necessary and drains power that could be used to lower your ETs.

The stock oil pump is good to at least 400 hp, and RPM of 6,500 to 7,000. Unless your motor will exceed this, it can be used stock unless you want to improve overall performance and cut power consumption. For engines within these limits, the stock pump's 45-psi oil pressure is all that's required.

Increasing the oil pressure just to be on the safe side may not prove as safe as you think, since it's possible to have too much oil pressure. Unnecessarily raising the oil pressure causes higher oil temperature and greater parasitic losses driving the pump. Look at it this way: the higher the oil pressure, the more backpressure there is to resist the turn on the pump. This means it takes more power to turn the pump. Not only is excess pressure an obstacle to power production, but also excess volume.

If the stock pump has adequate volume, installing a high-volume pump, which bypasses all the additional volume, serves only to reduce power output. Consider whether your application needs anything other than the stock pump. If it doesn't, don't spend your money on a higher-pressure higher-volume pump.

Although adequate for most applications, the stock pump can be improved. If you rework the oiling system, or just the main cap where the oil pump locates, the ease with which the oil discharges into the block will have been improved. This results in less pressure loss between pump and bearings.

Internal Pump Clearances

It's relatively minor here but if you want to get picky and have several pumps (used or otherwise) to choose from, a unit with tighter clearances can be built. The normal clearance between the gear OD and the case is between 0.002 and 0.004 inch. By selecting the housing and gears, a pump with 0.002-inch clearance instead of 0.004 can result. The gear-to-case end clearance is between 0.002 and 0.004 inch. If excessive, it can be reduced by lapping the end of the case on a flat surface. By reducing leakage, the pumping action is improved.

Oil pumps must be sized to satisfy low, not high, RPM requirements. A misconception is that engines need more oil as the RPM increases. However, the oil pressure/volume requirement isn't necessarily in proportion to the RPM. Bearing clearance does not significantly increase

with RPM, but the pump's output does. Unchecked, an oil pump's output increases far beyond requirements unless limited by the pressure relief valve. Consequently, all oil pumps have excess delivery at peak engine RPM. Because they are sized for the low-RPM applications, the oil-bypass system assumes a relatively important role. If bearing clearances are at the wide limit in an effort to cut bearing losses, tightening up the pump clearances helps idle oil pressure to the tune of 2 to 3 psi. At a 650-rpm idle, 10 psi is fine, but the pressure better be up to about 35 psi by the time the motor is turning 1,800 (or so) rpm.

Oil Pump Chatter

A characteristic of a small-block Chevy pump is a pulsating oil flow, which can cause pump chatter. This same vibrating motion also hits the distributor since its drive originates from the same point. This can cause a little spark scatter and inevitably lead to a small reduction in power output.

One way to reduce pump chatter involves the use of a die grinder. Remove the pump's top cap and inspect the discharge port in the cavity. On some pumps, during rotation, the driving gear intermittently covers part of the port. This causes the port's discharge area to change as much as 18 percent as the gear teeth go by. If the gears on your pump intermittently cover the discharge port, grind a lead-in with a bias directed toward the idler gear. Regardless of

By grinding the internal pump passages as seen here, the delivery of the pump is increased and power absorption reduced.

whether the gear covers the discharge port, it pays to streamline the entry because flow at the point where the gears mesh is disoriented.

Porting the discharge port not only increases the pump's volume of flow but also cuts the power needed to drive it.

A popular way to further reduce pump chatter on a stock seven-tooth small-block Chevy gear pump is to cut anti-chatter grooves into various parts of the body. How these grooves operate to overcome chatter has never been satisfactorily explained to me, so for want of a theory, I'll simply give you mine.

The anti-chatter grooves operate on two principles. First, they apply pressure to the end of the gears, which in turn presses them onto the end of the case for friction damping. Second, the grooves

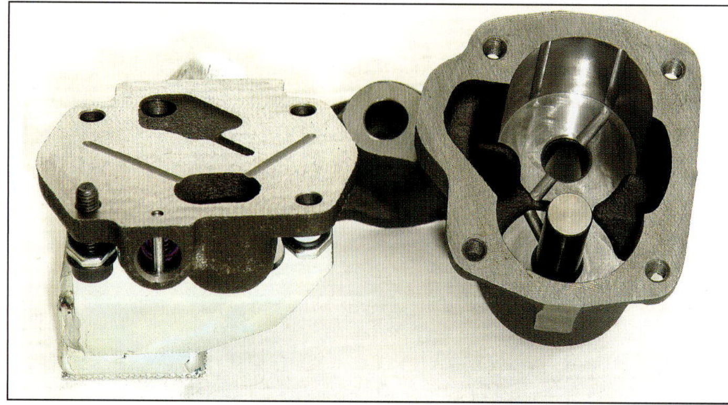

The Moroso pump shown here has anti-chatter grooves cut in it for smoother operation of the distributor.

spill some of the oil as it's squeezed from the meshing gear teeth, thus making the trapping pulse less severe. If this is the case, then any anti-chatter grooves may have the effect of reducing the pump's output. This, of course, won't be of consequence if the pump's output already exceeds requirements.

Pressure-Relief Valve

Most stock small-block Chevy pumps deliver oil pressure between 35 and 45 psi. Most are right around the 40-psi mark. On occasion, changing to synthetic oil on an engine with the pressure relief valve at the lower mark will cause the oil pressure to drop as much as another 5 psi. If a sagging relief valvespring is suspected, buy a new spring and compare the stiffness and length to the existing one.

If an increase in oil pressure is required, fit a stronger spring—Chevrolet P/N 3848911 for 55 to 60 psi—or shim up the existing spring. If you shim the spring, be careful that you don't limit the travel of the piston such that it will not bypass the oil adequately.

Oil Bypass

On stock small-block Chevys and most replacement pumps, the oil is bypassed internally. This means that if the oil pressure at the outlet side goes above the bypass valve setting, the bypass opens and returns the oil to the pump's intake side. At high RPM, a high bypass condition exists, which means some oil may circulate a number of times before being delivered to the bearings. This causes the oil to be unnecessarily heated.

Though the amount of heat put into the oil doesn't constitute a major problem, steps can be taken to reduce it while increasing pump efficiency. Within reason, the larger the return to the inlet side of the pump, the better. On most pumps this can be improved by taking out the press-in plug on the cross-drilling that connects the pressure relief valve drilling with the intake, and re-sizing it. You also can help improve the return flow by accessing this area with a die grinder and radiusing off the edge of the cross-drilled hole.

A rarely used alternative to dumping the bypassed oil back to the inlet side of the pump is to bypass it directly back into the pan. Although nothing complex is involved, stock pumps rarely are modified to do this. There's no difficulty modifying an existing pump if you have access to a few simple pieces of equipment. Essentially, it involves blocking off the original transfer hole between the pressure relief valve and the pick-up pipe drilling. When plugging the hole, make sure a totally air- and oil-tight fit is produced. If an air leak exists, it will compromise pump performance.

With the original bypass blocked, you need to consider how the bypassed oil enters the pan. Unless precautions are taken to the contrary, the opening pressure relief valve can send a high-pressure stream of oil straight into the pan. This can splash-aerate the oil in the pan considerably (we will see, in Chapter 7, how this can cost measurable horsepower), so it should be put through some kind of grid to remove some of its kinetic energy so that it drops into the pan and creates less splashing.

Another method is to couple up a pipe to the pressure relief valve passage and direct the oil toward the side of the pan. Adopting this technique achieves two things: it cuts the temperature of the oil, and it appears to markedly reduce pump chatter. This indicates that the pulsing bypassed oil is fed through the transfer passage back into the intake side is aggravating the system's tendency to pulse.

By externally bypassing oil, a problem can be created on the induction side of the pump as the demand through the pick-up pipe is increased greatly. This can cause the pump to cavitate and momentarily starve the bearings. To combat this, the stock 1/2-inch pickup, which at best is marginal, needs to be enlarged to about 5/8-inch diameter.

Alternative Oil Pumps

Though there are ways and means of up-rating a stock pump, there are also good reasons for using a pump with different characteristics. For instance, if bearing clearances have been increased, and the engine is using a low viscosity oil or high operating temperatures are expected, then the situation may warrant the use of a high-pressure, high-volume pump. We've already discussed the technique for increasing the pressure of the stock pump, which may be a wise move if high RPM is to be used and bearing clearances are to remain relatively stock. Under these conditions, 55 to 60 psi of oil pressure is desirable, but certainly no more than this.

One of the major manufacturers of oil pumps is Melling, whose products are widely available. Melling makes a stock replacement pump plus a high-volume pump (P/N 55HV), which delivers 27-percent more volume. This pump uses extended seven-tooth gears and is approximately 1/4-inch longer than stock.

Melling pumps are also equipped with stronger springs to increase the point at which the pressure-relief valve comes in. Springs available for the Melling pumps are color-coded according to stiffness. They come in plain, yellow, or pink, and their stiffness increases in that order. The plain spring sets the oil pressure at around 40 to 45 psi, the yellow one at about 55, and the pink one at 65 to 70 psi. Of course, these are hot oil pressures. When the oil is cold, pressures will be considerably higher.

If you're installing a replacement high-volume pump, be aware that these pumps are longer. If you use the stock pickup, it runs into the bottom of the stock pan, so an appropriate Melling pickup must be used.

Another alternative is to use a big-block pump for a regular small-block Chevy. The big-block pump has about 30-percent greater flow rate than the small-block pump, and it has a 12-tooth gear arrangement instead of the normal 7-tooth gear of the stock pump. Suitably prepped with an external bypass, this pump can reduce spark scatter substantially.

Oil Pan Design

When the small-block Chevy was originally designed, the pan was simply an oil reservoir. As performance levels of both chassis and engines increased, weaknesses were found in the stock design. The stock pan falls short in two areas: First, a car that's capable of accelerating or cornering rapidly causes most of the oil to move to one end of the pan and away from the pickup, resulting in loss of oil pressure. If bearings are to survive, it is vital to make sure the pump pick up is always immersed in oil under all conceivable operating conditions.

Second, the pan's innards must exercise some control over the oil to limit crankshaft entrainment. It's understandable to assume that since a crankshaft is rotating at a high speed, oil is centrifuged off. This is not the case. Granted, the rotating assembly centrifuges off a lot of oil, but much of the oil hits the crankcase or pan walls and bounces back into the rotating parts and becomes entrained. This can cost much more power than you may at first suspect.

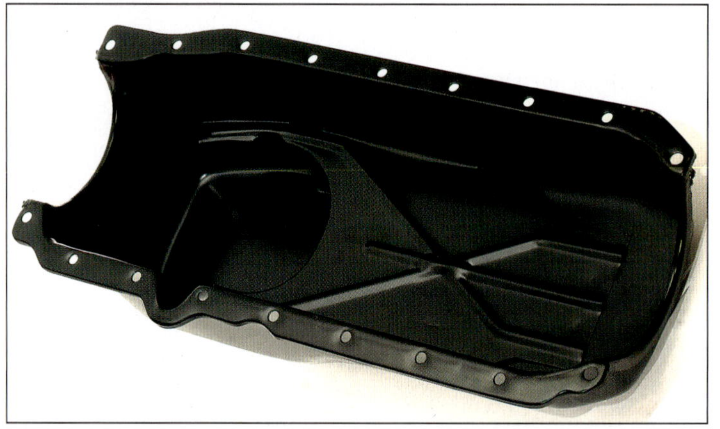

Here is a stock replacement pan from AAEQ. It has a plastic-like surface finish, a functional surge tray, and is really well made for the money.

Oil Pan Modifications

For a stock or near-stock motor, crank windage and oil entrainment in the crank may not appear to be of any great concern. However, some test results may convince you otherwise. Tests were made on a 400-ci equipped with a calibrated sight level in the drain plug hole. Oil was progressively added to the engine between power tests. At a little over a half-quart above the full mark, the crank began to gather oil.

Normal pan windage and oil dispersion through the rest of the engine usually causes oil level at the sight plug to drop about three quarters of a quart. With a half-quart overfill, 4,000 rpm proved the critical speed where the crankshaft oil entrainment became significant. At this RPM, the oil level observed at the sight plug dropped almost two quarts and 10 hp disap-

peared. Most street motors regularly turn 4,000+ rpm, so this is significant.

In this test, the overfilled pan replicated the situation produced by a vehicle under acceleration. Unless adequately baffled, the oil will migrate to the rear of the pan. A 1-g acceleration rate will cause the oil to be at 45 degrees, which means the rear of the crank is well and truly immersed. Just because only one part of the crankshaft starts to dip into the oil doesn't mean that entrainment stays local to that area. Under hard acceleration, the oil level rises at the rear of the pan, and the rear counterweight may start gathering oil. The most obvious conclusion is that oil will be entrained only at the rear of the crank.

In reality, because of the position of the counterweight, the crankshaft acts as a crude propeller and drives the oil forward. The oil then starts to form rope-like tentacles extending the length of the crankshaft. Even though the crank may only have dipped into the oil in one place, the windage and viscous shear problem occurs down the entire crank length.

The shorter the stroke of the engine the less likely there will be any oil entrainment. But we are not dealing with anything less than a 3.48 stroke here. Oil entrainment is a reality in the pan of a 3.48-stroke engine and it gets worse very rapidly as the stroke gets longer. For a 383, oil entrainment is a factor that needs to be seriously addressed. With modern tires capable of delivering good traction,

If you intend to build a stroker motor, this deep-style pan will accommodate up to 4 inches of stroke. Best yet this pan is not that much money!

and the fact that any small-block Chevy can turn over 6,000 rpm, the oil pan needs attention, even for minimal performance applications.

For street vehicles, the most obvious solution is a horizontal pan baffle to keep the oil from the crankshaft. If ground clearance problems don't exist, then use a deeper pan; the farther the oil reservoir is from the crankshaft, the better. If a deep pan cannot be accommodated, use a sheet metal baffle with appropriate drain-back holes to separate the oil from the spinning crank.

In an effort to reduce the amount of oil that remains on the walls, screens, and baffles, some companies coat the baffles and pan internals with PTFE. This encourages the oil to drain back into the reservoir area more quickly.

If a sheet metal windage tray is used, a scraper blade on the left-hand side of the block (as viewed from the front) should be used to shear off excess oil. The excess oil then should be directed underneath the windage tray into the reservoir below.

If the pan is shallow and there's little room for scrapers other than one on the block, a slightly different approach is necessary. If a solid windage tray is used, oil striking can bounce back into the rotating assembly. If a wire screen is used, it absorbs the oil's energy on impact. This allows it to pass through the screen rather than bouncing back into the crankshaft. Shallow pans then can benefit from the use of a mesh screen, whereas a deep pan and those with effective scrapers and/or recovery areas often work best with sheet metal screens.

Making an effective sump is quite a science, and although fabricating your own is not beyond the skills of anyone handy with a welding torch and sheet-metal cutters, guaranteeing top results is. I have had good results with pans from several companies but in the main I find that the Moroso range and quality of service well meets my needs.

For any serious competition where violent changes of direction are involved, this is the Moroso pan to use. Five trap doors control the oil surge and keep oil at the pickup. And just in case you are wondering, the parts bill for this engine was about $5,800—not bad for over 750 hp!

If you go online to Moroso.com and check out the pans available, you will see the range is very extensive, even for just the small-block Chevy. Take your time selecting here because you will need to take into account the size and style of the rear main seals (crank to block and block to pan) as well as the application. If you have any problems selecting, call Moroso's tech staff—this may save you returning an incorrect selection.

Top-End Oil Restriction

A significant proportion of the oil that finds its way into the crankshaft's domain originates from the valvetrain. Motors with the stock-type ball-mounted rockers need a good supply to the cylinder head components of the valvetrain to prevent the ball and rocker assembly from galling and overheating.

If roller rockers are used, the need for a large quantity of top-end oil is eliminated, as roller rockers require only a minimal amount of oil. Some oil will be necessary to cool the valvesprings, especially if they're doubles incorporating a flat-wound damper. The damper spring performs its func-

tion by generating friction between the two springs, and friction generates heat. Without sufficient oil to remove excessive heat, the valvesprings will quickly lose their temper and the closing force exerted will drop.

Oil for the lubrication of the upper end of the motor passes through the two lifter galleries, through the metering valve and the lifter, through the hollow stem of the pushrod, and out into the rocker. It drains back mostly via the drain-block holes at the ends of the cylinder heads, and from there it runs into the lifter valley. At this point the oil can pass back into the pan via several different routes. It can run out through the holes in the lifter valley, over the cam lobes, or down the drain-back holes at the front and back of the block.

If roller rockers are used, oil flow can be restricted at any number of points. Probably the best place to restrict the oil is at the lifter, but it's not always the most convenient. The most popular method is to install restrictors at the back of each of the oil galleries. This modification only applies to solid-lifter-type applications, because hydraulic lifters need the additional oil for their operation. Restricting

---

the oil at the back of the lifter galleries means there's a lot less oil returning to the crankcase from the top end and getting entrained in the rotating assembly.

If a roller cam is used, it's practical and desirable to plug the oil return holes at the center of the lifter valley. These holes usually serve to splash oil onto the camshaft for lobe lubrication, but with a roller follower this is unnecessary.

If a flat-tappet camshaft is used, it's possible to plug or put stand-off tubes in these holes, but only if relatively short mileages are intended. If you intend to use a flat-tappet cam on the street, don't plug them—it will lead to increased cam wear.

If the situation allows plugging of the holes at the center of the lifter valley (covered in Chapter 3), we have to assume that all the oil will return to the crankcase via either the two holes behind the timing chain or the drain-back holes at the back of the block. Generally, it's best to encourage the oil to drain back to the pan via the back of the block rather than from the front.

Oil Movement

For a drag-race engine, keeping the oil drain-back in the back rather than in the front is no problem, since acceleration forces it to the back of the block anyway.

For a road-race engine, the acceleration forces cause the oil to go down the back and, under braking conditions, down the front. Engine horsepower isn't needed during braking, so it's not super-critical.

In any event, it pays to make sure the block drains the oil down the face of the block, underneath the timing chain, and on to the pan as rapidly as possible. That's because getting back on the throttle may cause the oil that went through the front holes to be collected by the crank.

If roller rockers are being used, it pays to put a debris screen in each of the return holes in the block. Sometimes, a

Bores of my Cup Car motor after 1,000 race miles. Best is at top, worst at bottom. This is with JGR oil and ACES.

Whatever pan is used, be sure to use the appropriate oil pick-up.

rocker will break up, and when it does, needle bearings will pass into the lifter valley. If these get to the oil pump, the results are disastrous. Using epoxy resin, a screen can be installed in each of the four oil-return holes.

Here my crew-chief friend Mervyn Bonnett does the final tightening of the pan for our street stocker dirt car. Note the pan kick-out for turning to the left only. This class-legal $2,000 engine out-powered the fastest cars with engines of three times the cost by about a 30-hp margin. Traction though, proved to be a race-deciding factor—not power!

CYLINDER HEADS

A factor more influential toward successful high-performance engine building than any other is the cylinder heads. Without adequate airflow, the engine will never make power. The areas one inch before to about a half inch after the intake valve, and similarly for the exhaust, are the most difficult flow restrictions to minimize in the entire engine. In addition to flow, aspects such as swirl, port velocity, and combustion characteristics also play major parts. This means that the cylinder heads you elect to use and what you subsequently do to them is the prime factor dictating the power achieved.

Since the introduction of the small-block Chevy in 1955 up to the late 1990s, a small budget meant a limited choice of heads for performance. Until the turn of the millennia a restricted budget meant buying cylinder heads from a wrecking yard, a private third party source, a swap meet, or a reputable discount performance auto supplier. If a little luck swung our way we just might pick some functional aftermarket aluminum heads at a rock-bottom price on eBay. But luck is an element that can't be counted on.

Heads from Dart, EQ, and RHS can make big horsepower numbers right out of the box, and they don't cost an arm and a leg!

Fortunately, from about 2000 on, the amount we had to rely on luck started to decline considerably as far as heads were concerned. About that time, the aftermarket demand for low-cost performance orientated heads had grown to such an extent that affordable options expanded beyond almost everybody's expectations. When I wrote the first edition of this book, the budget meant we were almost certainly locked into using production heads. That is no longer the case. We have moved on, and I will deal only minimally with factory-produced heads. While on the subject, I will tell you what to avoid and what to use if your build involved factory heads. As far as aftermarket heads are concerned, there are several brands of iron heads that work well out of the box and port up very easily to deliver very professional results even when done by a novice. Also, volume production has meant that not only are aluminum heads a financial practicality, even when a relatively restricted budget is involved, but also at least, one brand of CNC head falls into an affordable budget for most engine builders. Sure it's at the top end of

---

the sort of budgets we are dealing with here, but nonetheless it is a good indicator of the expansion of the market that has taken place.

Let's deal with the options for those on the very smallest of budgets. Other than buying a set of used aftermarket heads that, of course, will limit us to factory heads.

Factory Heads For Performance

Although I have not tested every single small-block Chevy casting, I have pretty much tested every sub group. If your budget constrains you to use a factory casting, then be aware you can make or break your engine's final output right here. Here is what is needed in the way of heads for the best return on investment:
1. Good combustion characteristics.
2. Good flow potential.
3. A small enough chamber volume to get a working compression ratio without an excessively high, plug-masking, piston crown.

When making a choice of production heads, always consider these three points. If you lose sight of them, your project will suffer.

For the really-low-budget build, let's consider early iron heads—those cast before about 1980. Aside from cracking problems there are, for power production, two distinct groups of production heads. There are those that should be avoided like the plague because they don't make torque or horsepower, and there are those that do. Fortunately, it's easy to determine which group an unlisted cylinder head you may be looking at falls into. The combustion chamber style is usually the giveaway. Take a look at the combustion chamber photo on this page. This is the closed-style chamber as opposed

With early heads (pre-1973), this is the style of chamber that makes power. When milling these heads for compression, be sure to see that the groove (arrow) that circulates water around the plug boss is still deep enough to do so. If not, use a carbide cutter to deepen it.

to the "open" or smog-style chamber. The open chamber, used predominantly for low-compression car, truck, and smog applications from the mid-60s into the late 70s must be avoided. Using these open-chamber heads can cost up to 30 ft-lbs and a like amount of HP on a relatively mild street build. Avoid such castings even if they have the bigger 2.02/1.6 valves in them.

If the budget means early-style heads (as opposed to the later Vortec heads) then closed chamber heads are definitely the ones to look for. The only drawback to these heads is that they suffer from a little more valve shrouding (see Fig 6-1) than is necessary, and sometimes chamber cracking (details on cracking later in this chapter).

Once these heads have been overhauled porting is relatively straightforward in the initial stages and produces good results. However, getting ultimate results requires practice and a flow bench. Even in simple pocket-ported form and used in conjunction with other appropriate

parts, they can produce excellent results. The way to do both pocket and the more advanced porting is detailed later.

In addition to early heads, the 1986 and later Tuned Port Injected (TPI) 350 H.O. motors, such as the Pontiac Trans Am and Chevy Camaro, are worth having. These heads have chambers that look like they've inherited some smog-head characteristics, but they're sufficiently removed to fall well short of that category. They have relatively small combustion chambers, and the basic engine sports compression ratios of 9:1 or more. These heads will port out well, but be aware that 305 heads have smaller valves, so only look for 350 heads unless you're specifically working on a 305. This will save you the expense of extra machining to put in larger valves. Also, if the porting technician is too exuberant, there's the possibility of porting these into the water jacket.

The last two head styles on the list may be a little expensive, but they're increasingly becoming available on the used market. The first of these is the aluminum L98 Corvette head casting, produced between about 1986 and 1993. In stock form these heads are nothing special—just light.

Although they won't accept valves larger than 2.00 on the intake side and 1.55 on the exhaust without fitting larger inserts, they nonetheless can, in ported form as detailed later, produce some truly impressive results for a high-output street motor. On a 383 motor equipped with a single four-barrel carb, not only can the aluminum Corvette heads be made to produce in excess of 450 hp, but also deliver excellent low-end output for true street drivability.

The L31 heads from the late-model 350 Vortec engine,

Early Production Iron Performance Heads

Casting No. 3782461

The 461 castings were the first "double hump." They were produced from 1961 to 1966 and were installed on fuel injected Corvettes where they picked up the "fuelie" moniker. Made in both large and small valve forms, this casting has one of the lowest spark plug locations of any production head. A small batch of special castings with the number 461X were made for use in stock category drag racing, but were said to have been installed, in many instances, on trucks. The intake ports on these heads are enlarged from 158 cc to 175 cc, with a view to making more top end power. The chamber has a quench pad adjacent to the spark plug whereas all other heads are chamfered off in this area.

Casting No. 3890 462

As with many heads in this group, a double hump was used to identify the 462 head. The spark plug is located a little higher than on the 461s and, as is most common, the quench pad behind the plug is relieved. This is the most common style of chamber within the group of heads we are considering here. Valve sizes could be either large or small.

Casting No. 3917291

A large double hump marking is used on the 291 casting and it is virtually identical to the 462 casting. It was used predominantly on 1968 model year vehicles.

Casting No. 040

This casting appears to be the same as the 291 or 492 castings. It is identified by a triangle marking.

Casting No. 3927186

The 186 castings were produced mostly between 1968 and 1972. They were used extensively on 302, 327, and 350 engines. The ports and chambers are virtually identical to those on the 291 and 462 castings. These castings, unlike some others, have accessory bolt holes on the end and this has made them a popular choice on performance street applications to the extent that they are now scarce. One important aspect is that the water jacket coring is different from the 291/462 heads, leaving less thickness below the spring pads. As with other double hump castings, these heads came in large and small valve versions.

Casting No. 3947041

Used on 350s in 1969 through 1970, the 041 casting is per the 186, but usually thicker under the spring platforms. It can be identified by a triangle on the end face.

Casting No. 3991492

Installed on vehicles possibly as early as 1969, the 492 heads were used until 1971. This head casting is supposedly the refined version of the 186/041 casting, and was intended for high-performance, production-line vehicles. It is also a service replacement head for all 64-cc head vehicles, and can be had under PN 3958603 with press-in studs and PN 3987376 with screw-in studs. Heads produced before 1972 had casting cores similar to other production heads, but those produced later can be machined to take a 1.44-inch diameter spring.

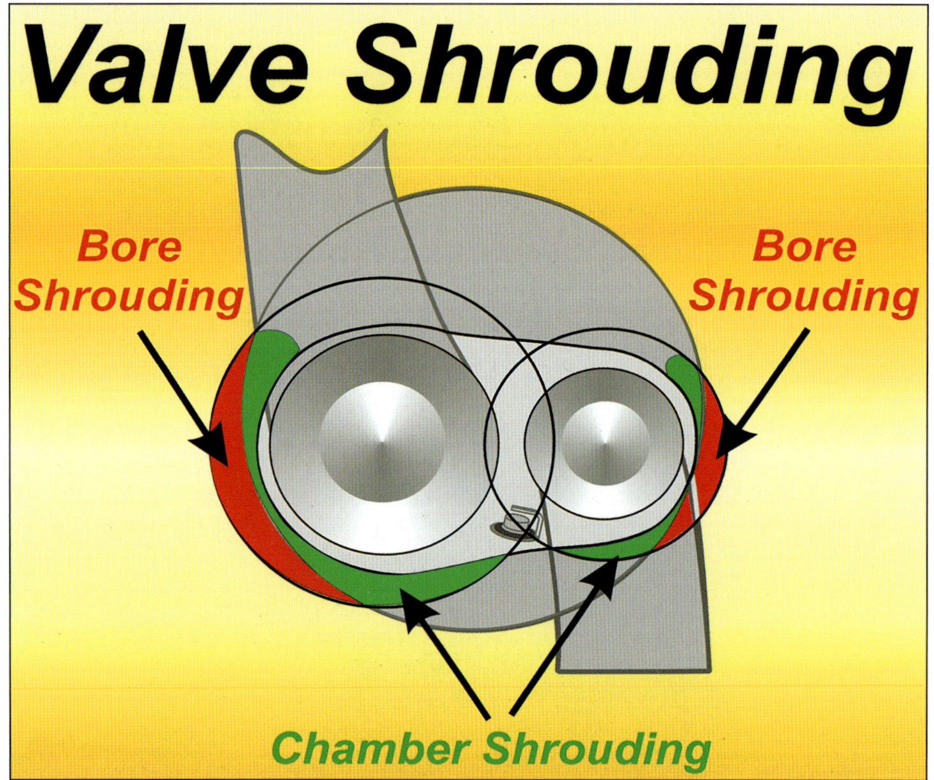

Fig 6-1. Valve shrouding caused by the close proximity of the bore walls (red shading) or the combustion chamber (green shading). If the casting is thick enough, combustion chamber shrouding can be dramatically reduced. The down side is the chamber volume gets bigger and may end up too big for the CR required—especially if a stock CID is involved.

For a valve to have geometrically unrestricted flow around its entire circumference during the opening events, it must have a clearance at least equal to the lift. This applies until the valve reaches one quarter of its diameter in lift (0.25D lift). The circle around the valves shows the room needed for it to be free of shrouding. This illustration also shows that a substantial amount of shrouding is caused by the cylinder wall. This we can do nothing about, as it's a consequence of a parallel two-valve cylinder head design.

This situation represents geometric shrouding and assumes that the air flows evenly around the valve. Unfortunately, this doesn't happen. The air's direction of approach to, or from, the valve affects the way in which it leaves the valve. In a small-block Chevy intake port, most of the air is flowing on the long side, so shrouding of the long side has more of a negative influence than shrouding on the short side. Shrouding on the long side, in the area toward the intake valve side of the spark plug, needs to be minimized, whereas shrouding on the short side requires only minimal attention.

introduced early in 1995, is the other type of cylinder head you need to look for on the used market. This is probably the best cylinder head casting that Chevrolet has ever produced for a production small-block Chevy. The casting quality is as good as the best aftermarket iron heads, and its flow capability on a valve-size-for-size basis more than matches the Phase 6 aluminum Bowtie heads.

With the installation of larger valves, it will not only generate sufficient flow for good top-end horsepower, but the Vortec's high swirl and port velocity also means exceptionally good low-end output. With just bigger valves and a simple pocket porting job, those with relatively little porting experience achieved favorable results (shown later). Therefore, you should put late-model Vortec heads high on your used, factory-head priority list.

Production Head Overhaul

Other than Vortec heads, most of the "desirable" castings you will locate could be as much as 30 years old, or older, and will need a rebuild. As stated earlier, you should have bought them with some kind of assurance that if they're cracked you can get your money back.

You will need to check the following and figure whatever it costs to do could have been money spent toward new aftermarket heads.

Head Preparation
- Crack test
- Oven bake/ball peen clean
- Check guide wear; new guides/guide job are almost a certainty here
- Machine for screw-in studs if springs heavier than stock are to be used
- Machine stud bosses for pushrod guide plates
- Valve seat machining
- Spring pocket machining for larger diameter springs
- Mill head face

Now your heads may not need all these ops done (although they most likely will), but even so, you are going to come out of this with a sizable machine shop bill. This means you should factor this in the cash equation. It may well be worth the effort to round up that extra money for a set of new aftermarket heads with all the features we are trying to achieve here, ready to go.

Guide Inspection

Let's talk a little more about guides before moving on. Ideally, the valve-stem-to-guide clearances need to be about 0.0015 inch on the intake and 0.002 inch for the exhaust clearance. With older used heads, this is unlikely to be the case, so let's look at the maximum limits. Anything more than 0.004 inch on the exhaust and 0.0035 inch on the intake will start to reduce power because the valves will not seat properly. Also, loose guides will cause whatever valve job may be done on the heads to wear faster because the valve lands on the seat in a different position each time.

Worn valve guides cost power. Position the valve as shown here and check side play. If it exceeds 0.015 inch the guide is worn too much to be used.

Fig 6-2. Your engine needs good valve guides for high performance. Here's why: at intake valve lifts above 0.500 inch, flow starts to establish a pattern different to the one that existed prior to this point. Instead of flowing out wherever minimal shrouding allows, the flow is directed across the back of the valve out into the center of the cylinder. By leaning or biasing the port accordingly, flow in the high-lift range can be enhanced when this flow regime comes into play. Without good guides, side loads from high lift will cause very rapid wear.

Worn guided can be fixed with oversize stem valves or the installation of new press-in guides.

Valve Sizes and Seat Recession

Decisions on valve sizes need to be made at this point. Anything less than 1.94/1.5 intake/exhaust combination in a bore of 4 inches or more isn't going to produce the most desirable results. The most common high-performance valve size combination is 2.02 inch for the intake and 1.6 for the exhaust.

If your existing valves have good stems, the use of 1.94 intakes doesn't give away significantly to 2.02s. If wear dictates going to the 2.02s, then have the seats and chamber cut accordingly. The installation of the bigger valves can cost as much as $50 over that of a regular valve job. However, be aware that if you decide to stick to the 1.94/1.5 valve combination, you still can get decent results at a lower cost.

As far as a valve seat job is concerned the usual situation is that the valves, especially the exhaust, have become recessed into the head. Although machine shops offer a shallow-angle valve seat top-cut service that removes the effect of small or moderate amounts of recession, it still can leave the valve too low in the head. If the cylinder head looks as if it needs work to cure recession, consider having larger valves installed because that has the potential for more flow as well as fixing the recession problem.

Another factor to consider is that earlier production heads don't have armored seats for use with unleaded fuel. Don't worry about this, unless your intent is to build a high-mileage street driver. If you must have armored seats, you should consider using an aftermarket head because these heads have them out of the box.

Most beginners wrongly presume that port shape and roughness are the flow-limiting factors. All too often it's assumed that the valve seats play only a minor role as far as flow is concerned. It's certainly appreciated that the seat plays its part at low lift, but the significance doesn't stop at low lift, as we shall see later. Take a look at Fig 6-3 and Fig 6-4 for some working dimensions.

Correcting Excessive Guide Clearance

If seats need reconditioning, consider the method of acquiring the desired valve-stem-to-guide clearance. It is most cost effective to use valves that have oversized stems. PEP and Engine Tech (and others) supply such valves with oversized stems (available at many engine re-con shops). Engine Tech also supplies valves that not only have oversized stems, but also heads typically 0.025- to 0.030-inch oversized. This is an ideal, low-cost way to fix recession in

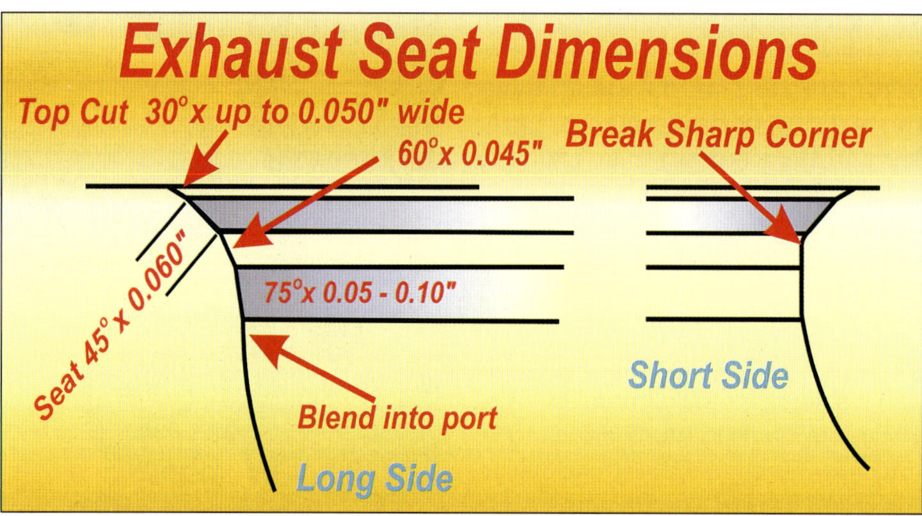

Fig 6-3. The exhaust seats cut as shown here work well on factory production-line iron castings made prior to the advent of the Vortec heads. Note the form used on the side adjacent to the short side turn.

heads that already have the 2.02/1.6 combination, as do some production line, high-performance vehicles.

Cast-iron guides, though functional in terms of wear, aren't necessarily the best. I recommend two options: the thin-wall bronze guides, such as those produced by K-Line or PEP, or the thick-wall bronze guides commonly available from most engine re-con shops. All are effective in terms of low wear rates.

If you're putting together a cost-conscious head package and you want the best bang for the buck, be aware you are going to need to do some checking around. Many machine shops are not eager to take on performance-orientated work. Now that may sound like bad news but on the flip side of the coin, the high-performance market is expanding while the plain engine reconditioning business is shrinking. The shops that are surviving this contraction are those willing to do work at a reasonable cost for folks like us who want to go as fast as our slim budgets will allow. To get the most from a basic 3-angle valve job, the best plan is to photocopy our seat drawings and ask what it will cost to get your seats done like that. It will vary from head to head as the amount of metal to come out will be different depending on what size of valves where there to start with.

Along with a valve seat job, some basic valve reshaping will be required to make it all work well together. Almost all the valves available off-the-shelf need some work done to improve airflow.

There are many good brands but this is the type of valve stem seal (right) needed for the typical bronze guides seen here.

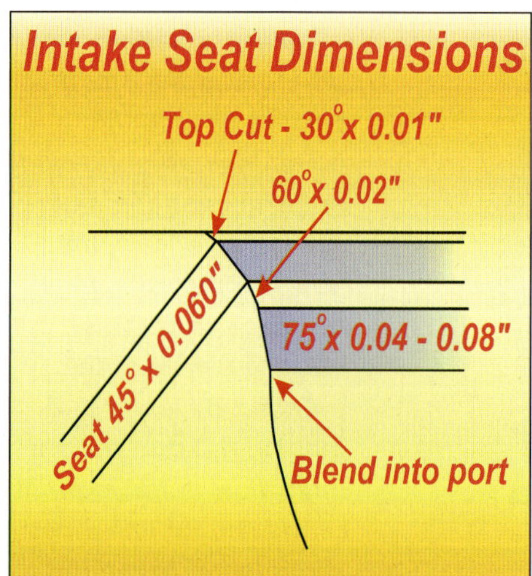

Fig 6-4. When rebuilding a set of heads, intake seats cut to this form work well and don't usually cost too much to have done.

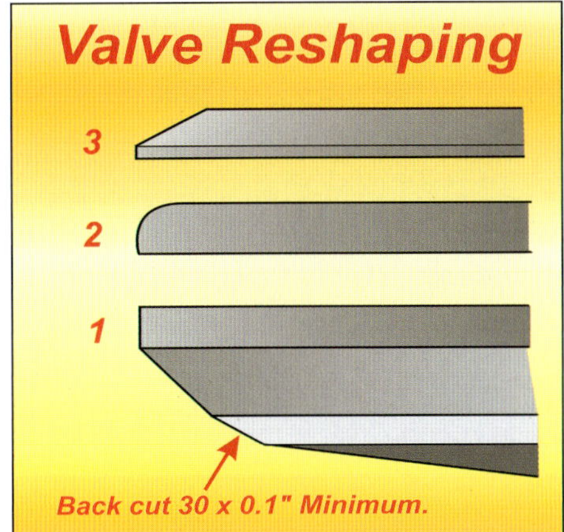

Fig 6-5. The valve land form of an out-of-the-box valve is typically as depicted by the shape No. 1. If a cam of less than 270 degrees of seat duration is used, both the intake and exhaust lands can be shaped as per No. 2. If the cam is bigger than 270 degrees, then the intake only should be shaped as per No. 3.

The simplest and most effective mod is to "back cut" both the intake and exhaust valves as per Fig 6-5. The final valve seat width needs to be just a little more than that used in the head. On the exhaust valve, it's important that the front face-to-margin has a generous radius. This radius considerably enhances low-lift flow and cuts the temperature of the exhaust valve.

Accommodating High-Performance Springs

The next few moves with production-line heads depend on cam choice (see Chapter 7) and the springs it requires. If the valve spring loads are not going to exceed 220 pounds at full lift, you can, for the most part, get away with the stock press-in studs. If the intended spring, cam, and spring combination to be used have sufficient lift to warrant a spring of significantly more than 220 pounds over the nose, then screw-in

Make sure all your valveseats are accurately done by check-lapping them with very fine oil-based lapping paste.

studs are needed. If these are a requirement, you will almost certainly need to convert to guide plates for the pushrod instead of the plain slot in the casting. This is easily done at the time of stud installation.

The more aggressive cams will require the spring pocket machined to accept a larger-diameter spring. This operation costs money even if you buy the special tool from the

cam company and, with an electric drill, do it yourself. So when you're selecting a camshaft, you need to consider what spring it should be used with and whether this spring requires the heads to be machined and whether or not the head can be machined sufficiently for the spring intended.

Basic Small-Block Chevy Head Porting

The following information on porting deals with 23-degree Chevy heads pretty much across the board. Applying what is detailed in the next dozen pages on heads will allow you to generate more airflow and, consequently, more power.

But just a moment, you're a beginner and you don't have a flow bench. If you fall into that category, listen carefully. The most important part of your cylinder head is, as per Fig 6-3 and Fig 6-4, the area 1 inch before the valve seat to 1/2 inch after it on the intake and vice versa on the exhaust. Be aware that what you think is needed may not be what is wanted. There are a lot of little wrinkles and nuances in porting that only experience and frequent use of a flow bench can teach. If you are using

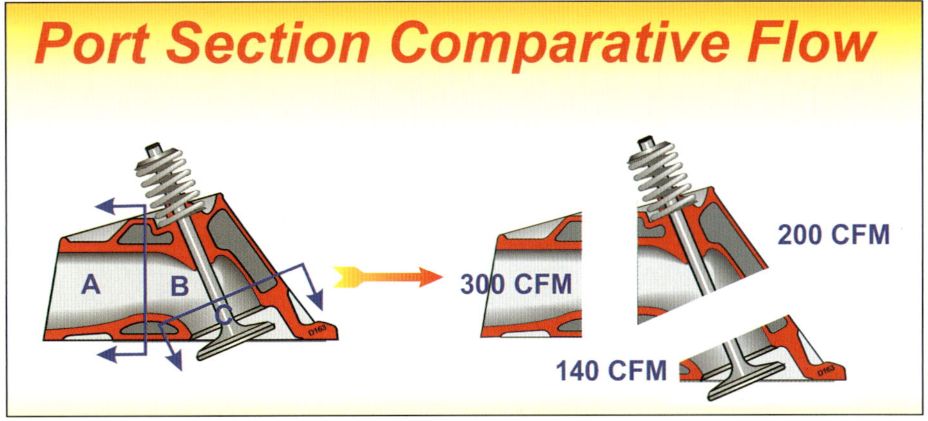

This breakdown of the port shows why grinding on the port entrance in an attempt to increase flow is the wrong place to start. The valve and the area 1-inch before to 1/2-inch after is the most important as it restricts the flow the most.

factory production-line heads and you do not, or cannot, spend a lot of time porting them, the good news is that there is an option.

Pocket porting and attention to the short-side turn will, while sensible street cams are employed, net about 80 to 85 percent of the potential power increase possible. Only when the valvetrain lifts the valve significantly above the 0.500-inch mark will a full porting job show big (as opposed to small or moderate) increases over a pocket porting job.

As the flow curves on these these pages show, the effect of extensive porting only pays off significantly at lifts above 0.450 to 0.500 inch. Prior to that, the seats and the valve pockets/short-side turn have the greatest influence.

It's tempting to carve out the space between the pushrod holes as soon as you get hold of the grinder. However, you should consider that this section of the port is already a straight shot into the cylinder head. It's close to 100 percent efficient, and until the valve's flow is substantially improved, the pushrod pinch point is not a flow impediment. Sure, at full valve lift it may cause a little flow loss, but remember the valve is only at full lift once in a cycle. It's at half lift twice, so what happens at half lift usually is more important.

In addition to flow efficiency, swirl is also important, especially for a good all-around street performer. Swirl can be generated or lost depending on how the port is shaped. The first point to be aware of is that virtually all production small-block Chevy ports have "port bias" in the throat or bowl area. Fig 6-2 shows the bias angle of a typical small-block Chevy intake port. Don't attempt to straighten this out. Instead, grind the port to emphasize this bias.

Another factor to consider is that the guide and valve stem take up room in the port and effectively cut the flow area. To offset this, the port needs to be widened around the base of the guide boss at the bottom of the port (Fig. 6-6). Applying the techniques described here to a full-race port spec typically takes about 100 hours. For a set of heads like this, when used on a 350 with an 11:1 CR, the best I have seen on the dyno is 518 hp and 462 ft-lbs. The cam used here was a short race Comp Cams single-pattern 285-2 solid roller with the lobes on a 108-lobe centerline. Given a 310 seat duration cam about 0.650 lift and a 14.5:1 compression, a set of heads like this should allow a ported Victor Jr–equipped 355 to make about 560 hp.

But as good as all the foregoing may be, let us not forget the amount of work involved. A hundred hours can only be justified if there are no other castings available. But there are. Remember that Vortec heads have been around since late 1995. These will port up in a third of the time and show better results to boot. Then, there are the aftermarket heads that we will deal with later. The ones I have chosen to show this time around will port up in less than 15 hours and, with experience, some in less than 8, while producing not just good, but truly outstanding results.

Port Re-Shaping Techniques

At this point, we can start to look at the actual techniques necessary to rework ports and chambers. First, in almost all circumstances except the combustion chamber, the surface finish should not have a high polish. For the most part, you need only use one or another of the carbide cutters shown in the "Porting Supplies" sidebar on page 70. Any of these plus a steady hand can be used to do 90

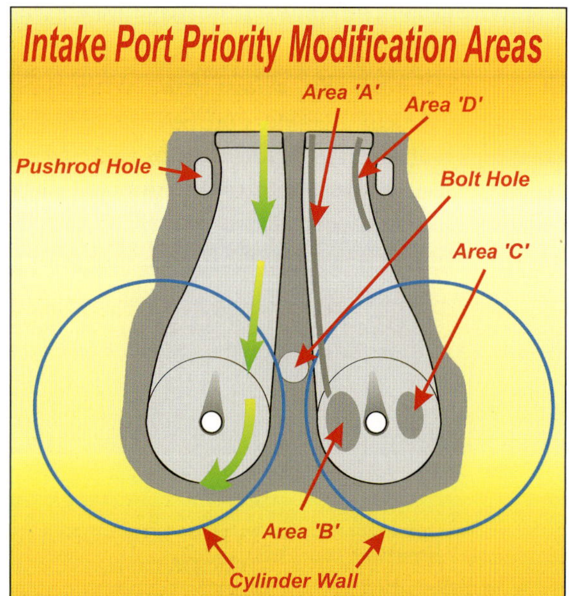

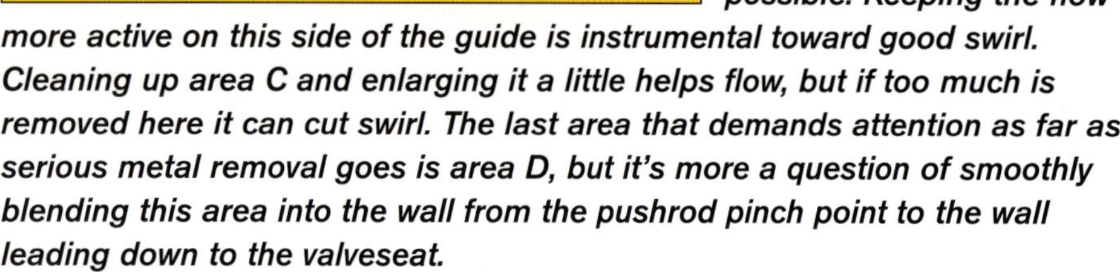

Fig 6-6. The left hand port shows the principal flow path into the cylinder. Because the flow is predominantly on the outer wall (A) and the port roof, we should give the greatest priority to these areas. As the air stream approaches the bowl, it encounters the guide, which offers measurable impediment to flow unless appropriately dealt with. The technique here is to raise and widen the roof at area B and, in the process slim the guide boss as much as possible. Keeping the flow more active on this side of the guide is instrumental toward good swirl. Cleaning up area C and enlarging it a little helps flow, but if too much is removed here it can cut swirl. The last area that demands attention as far as serious metal removal goes is area D, but it's more a question of smoothly blending this area into the wall from the pushrod pinch point to the wall leading down to the valveseat.

percent of the work on the heads. Why? The form accounts for 98 percent of the success of a port's capability. What you may find, though, is that controlling a carbide to get the final form is a little beyond the capability of a novice. This is where the cartridge rolls come in. A coarse-grit roll, say 80, or even 60 grit, can be used more effectively to achieve the final forms being sought.

However, for novices, a point of concern is blending the ports and chambers into the valve seat areas. On the chamber side, the way to tackle this is to use an old valve and cut it down. Working on the port below the valve seat presents a similar, if less difficult, problem. Painting the seat and the area on down into the port a little way with engineer's layout blue helps. When grinding in the throat avoid getting any closer to the lower edge of the seat than 1/16 inch.

Once all the cutting and grinding is done, the finishing touches can be applied using an emery roll. These can be had, at a very competitive cost, from Dr. Air, as discussed in the sidebar "Porting Supplies" on page 70. Normally, you will use three or maybe four 80-grit emery rolls per cylinder. To economize on rolls, do essential work first, and if they are not worn out use the rolls to do the finer finishing work. This usually means doing the seat area first and then working down into the port or up into the chamber. Polishing the chamber to a relatively fine finish is a good idea. Having a fine finish here cuts heat conductivity and reduces detonation-inducing hot spots.

Valves

Understanding that the seats in the head are important is only half the battle. The shape of the valves is equally important. With stock valves, there is not a lot we can do except basic preparation. This involves blending the back face of the valve into the seat on both the intake and exhaust. On the exhaust valve, blend the top face into the margin using a generous radius. Fig 6-5 shows what is needed. Don't skip this operation!

Exhaust Ports

Inspection of the exhaust ports reveals one port in each head (except aluminum Corvette heads) has a heat riser cross-over passage that communicates exhaust to the heat passage in the intake manifold. This, in terms of flow, looks pretty ugly but, in reality, has little affect on flow. The exhaust valve seat, for these early cast-iron factory heads, should be cut according to Fig 6-7. This is a simple form

that works well and can be done by most machine shops with even the most basic seat-cutting equipment. If we are porting Vortec or aluminum Corvette heads, an exhaust seat with a generous radius under it to blend into the rest of the port is the way to go. Exhaust seats/ports like this are used on most high-performance aftermarket heads. This offers a distinct advantage if the rest of the port, especially in the guide boss area, is shaped to work in conjunction with a radius design.

Three factors are important on the exhaust port, especially when only pocket porting is being done: the short-side turn, the port bias (as this strongly influences the high lift flow), and the area at the base of the guide boss.

The exit at the manifold face is not a critical area until a refined state of port development is reached. Just as with the intake, the actual body of the exhaust port can come off the short-side-turn seat

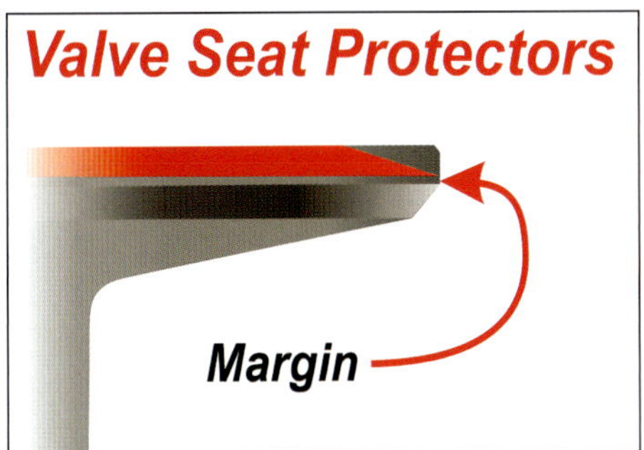

Grind or turn down the top face of the valve as shown here. Leave a margin between the angle on the top face and the valveseat of about 0.015 inch. This will make the cutter or emery roll stop short of the actual valveseat and preserve most of the width of the top cut.

Although the mods shown on these pages will get good results without the aid of a flow bench you may still want one—but cannot afford it. If so you can get full details on how to build one delivering professional results—in a weekend—for less than $200 on MotorTecmagazine.com.

The first step toward establishing the CR your engine might have is to measure the combustion chamber volume. The Power House division of COMP Cams has inexpensive burettes and chamber plates for this job.

Porting Supplies

A good source of porting supplies is swap meets. It's been my experience that a wider range of metal-cutting and finishing products—with significantly better prices—are found at swap meets rather than at regular tool-supply outlets.

Most of the metal removal you're likely to do will require the use of carbide cutters. Not only can carbides be expensive if you're buying them in ones and twos, but also the type needed often depends upon the application. Although cast iron isn't fussy, aluminum can be. Also, we have to remember that it is porting we're doing, which often requires a longer-than-normal carbide shank. Along with that, more specialized tooth forms and head shapes are advantageous.

Lastly, cutting fluids are required for aluminum work, which prevent the cutter teeth from loading up and reducing the cutting action to almost zero, are a very useful aid to speed production as well as surface finish. With the right cutting agent, a cutter finish is almost good enough for most applications. The sample bottle of cutting fluid in the Dr. Air starter kit, as shown here, is considered by professionals to be about the best available. Small though it may be, the sample bottle is good for probably a dozen pairs of aluminum heads.

When it comes to porting supplies there are few tool outlets that actually have any of the specialist stuff that a serious porting exercise might call for. Here I strongly recommend Dr. Air brand porting supplies. These are porting products that, with cooperation of the manufacturers, have been developed or refined over the years by Roger "Dr. Air" Helgesen. This has resulted in cutters that remove metal faster and produce a better finish. The Dr. Air cutting fluid, a sample bottle of which is supplied in the "8+ cylinder kit" (which, unlike most kits you can buy, actually does port eight or more cylinder-head cylinders)

is considered to be an industry standard and is used by many top pro head porters. Along with this, the emery rolls have more aggressive cutting characteristics and last longer than most competitive brands.

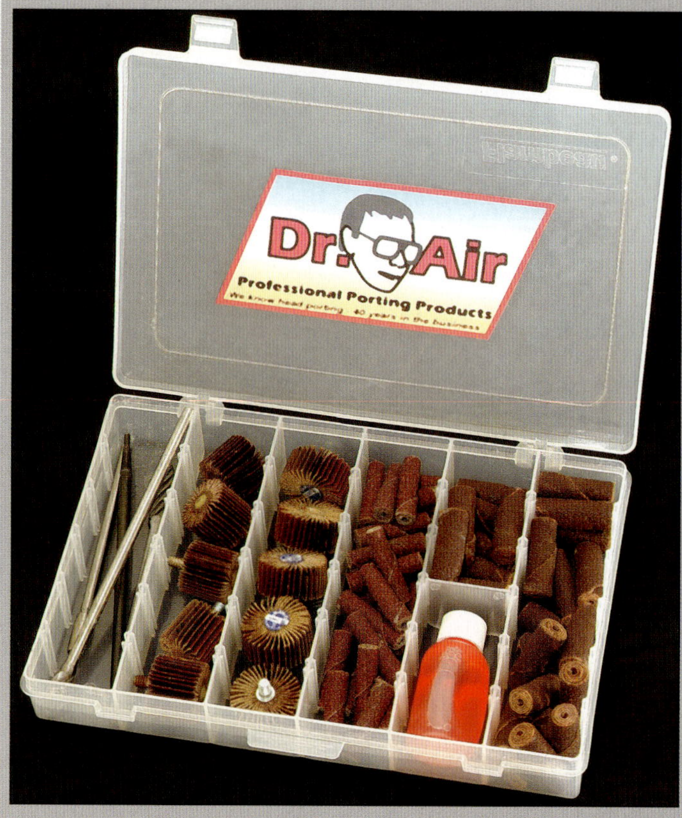

The last and most important aspect is the cost of these supplies. The cutters are very competitively priced for a product that is top quality and the cutters supplied in the starter kit will cover about 99 percent of most home head porters needs. The best part of the Dr. Air porting supplies is the emery rolls and the like. First, they are close to half the price of what you may be asked to pay for other leading brands. Second, they are packed in boxes of 50. This is a far better deal for those who are porting heads at no more than, say, a pair every few months or so, than much more expensive boxes of 100. But that is not the last point on cost. The Dr. Air 50 Pack Bonus Box is so called because they claim not to have enough time to count out exactly 50 rolls per box so they throw in a few more to make sure there are at least 50 in the box. This averages out at about 53 per box, hence the term "bonus box."

Dr. Air porting supplies are available exclusively from the California-based company of (coincidently) Dr. J's Performance. They can be contacted at 714-943-3404 or you can check out their online catalog at J-performance.com.

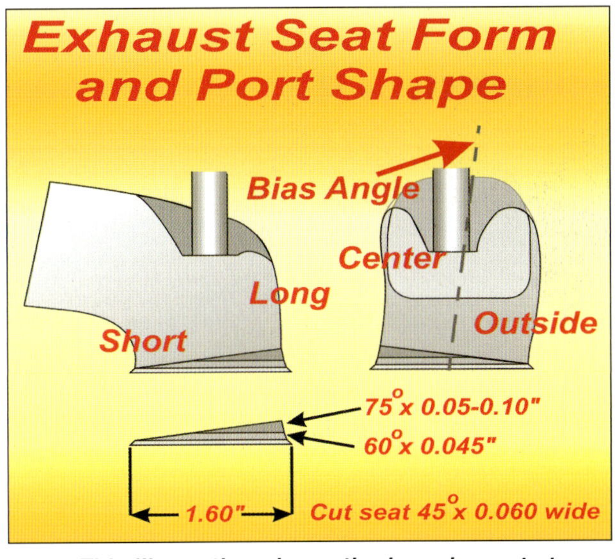

Fig. 6-7. This illustration shows the key elements toward finding exhaust flow and with usually little machine shop expense. Note the bias angle in the right-hand drawing, and that the port's outside wall is nearest the cylinder wall. The port-to-seat junction on the outside of the port wall doesn't need a blending form of any consequence, but the port wall and seat on the center does, as this is a high-flow area. Also because the main body of flow is on the long side, it is important that the port-to-seat junction (left-hand drawing) has a generous blending form.

area more abruptly than it does on the long side on the long-side turn. If the short-side area of the seat flows too much, it causes a high-speed-flow interference issue on the long side. By deliberately reducing the flow efficiency at the seat on the short side, the amount of exhaust exiting here at the mid- and high-lift is reduced so the long side is allowed to work much better. However, once immediately past, the seat area shaping of the short-side port floor significantly influences bulk flow of the exhaust port. If you stick to the form shown in Fig 6-7 things will work just fine.

Compression: The Budget Racer's Big-Power Ticket

Pay attention here, or it will be the downfall of your quest for best torque and HP from your engine. Once the minor to all-out porting work is done, it will be time to consider what combustion chamber volume will be required to achieve the desire compression ratio. Before arbitrarily deciding on the compression ratio your engine might need check the cam chapter (Chapter 7). From what is discussed there a more informed decision on the cam and compression ratio can be made.

Here, an explanation on achieving cam event and compression compatibility is warranted. Taking into account the fuel octane to be used there is a certain minimum compression ratio (CR) for any given cam duration if the best torque and HP per cube is to be achieved. If the CR drops below this minimum, it becomes better to choose a slightly shorter cam as the combination of a shorter cam with a slightly-lower-than-optimal CR being used will produce a better power curve than a longer cam with the same CR.

This is not so noticeable with moderate CRs and cams in the 250- to 275-degree (off the seat duration) range, but as the cams start to fall into the racier category, it becomes very important. This factor is one of those parts compatibility things so often alluded to but rarely defined.

Once a cam decision has been made, you will need to establish just what it will take to get that ratio. Here I should point out that, assuming the fuel octane is there, having a CR higher than the cam spec calls for is good, but having it lower is not. In other words, assuming detonation is not a factor, there is hardly a situation where too much CR is a problem.

Although a dial caliper can be used, these spring height micrometers (COMP Cams) are not that much money and make the job easier.

After consulting Chapter 7 to establish what CR is needed, it's time to establish what it will take to get it. The first move is to establish the existing CR. To do this, you need to measure the combustion chamber (in cubic centimeters). You also need to make a provisional choice of piston in terms of crown configuration. Remember, a flat-top piston has many advantages in terms of results, weight, and cost. If the CR you are shooting for needs to have a raised-crown piston, go for the minimum possible. The best combination is almost always achieved with the smallest chamber in the heads and the least amount of piston-crown rise. Using too much of a raised crown on the piston together with many early-style heads with the low plug placement can mask the plugs. This can really inhibit the combustion process and the result is a poor output.

Since getting an adequate CR is sometimes a problem, adding cubes is a great help toward making more compression with a flat-top piston. It costs money to mill heads to get compression, and money spent here would go about 50 percent of the way toward the cost of a longer stroke crank that, often, will ultimately do the same thing for the CR.

All the foregoing leaves us having to machine our factory production cylinder heads as much as possible to minimize chamber volume.

Unfortunately, there is a milling limit imposed by the proximity of the intake valve seat and the cylinder-head face being milled. As the head is milled for compression the face gets closer to the intake seat until the face and seat actually meet. At this point, further milling also cuts into the valve seat. Take a look at your heads, and you will see that it won't take much in the way of machining for the head face to cut into the valve seat on the shallow side of the chamber. Depending on the position of the seats, the coming together of the head face and seat usually occurs, on stock heads, when 0.030 to 0.050 inch is milled from the head face.

With closed-chamber heads, every 0.010 inch machined off drops the chamber volume by 1.3 to 1.4 cc. Flat milling (as conventional head milling is called) of heads then leaves us with chambers at best 4.5 to 7 cc smaller. What this means is that flat milling usually only gets the heads to the 60-cc chamber volume point. Some heads get machined as low as 58 cc, but don't count on it. Volumes around the 60-cc mark are fine for a sane low-buck street 350 with flat-top pistons in flush to the top of the block. This, with a 0.038 Fel-Pro gasket and typical 4.8-cc valve cutouts in the pistons, will net between 10.4 and 10.6:1 CR.

If more compression, say 12.5:1, is needed for use with a race cam, then the total chamber volume (including head gasket volume etc.) needs to be reduced by a further 12 cc. Let me tell you now that won't happen, but we can, by angle milling

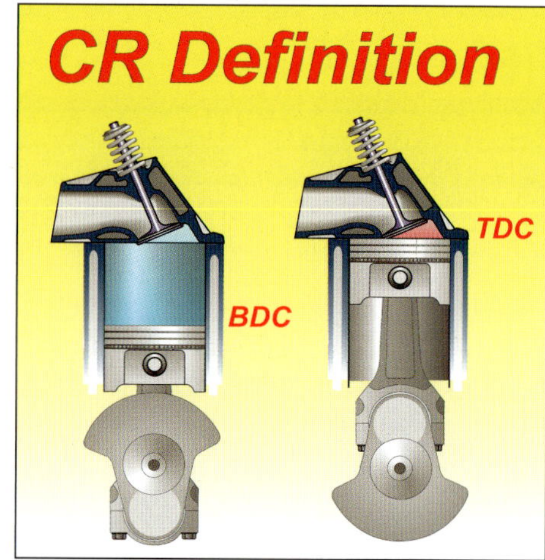

Fig 6-8. Here we see exactly what the term "Compression Ratio" defines. Essentially it is the total volume above the piston when the piston is at Bottom Dead Center (BDC) as shown on the left, divided by the volume above the piston at Top Dead Center (TDC) as shown on the right. The formula to calculate the CR is: CR=(V+C)/C. V is the swept volume of the cylinder in cc's and C is the total combustion chamber space. This not only includes the chamber volume in the head but also the gasket volume and any volume contained in the block when the piston is at TDC as well as any piston dish volume.*

the heads, get more than halfway there.

Angle milling is our alternative to conventional flat milling. By angling the heads at about 1 to 1.5 degrees, virtually zero metal is taken off the intake port side of the head and a maximum on the exhaust port side. By angle milling, the chamber can be reduced in volume to a far greater extent. Ultimately, the limit to angle milling the heads is set by two factors: the thickness of the outside edge of the casting (1/8-inch minimum) and the head face meeting the spark plug holes.

I have cut 186-style casting and achieved 53-cc chambers for an

11.7:1 CR with flat-top pistons, but be aware that heads machined this much are prone to cracking.

Angle milling raises the question as to whether we can do this operation without re-drilling or re-spot facing the bolt holes to bring them back true to the head face. For the amount we're going to angle-cut the heads, the answer is that it's best to do so but not 100 percent necessary. Another potential hang-up with angle milling is that the longer bolts on the inside of the heads can hang up on the side of the bolt holes toward the top of the casting. This interference must be remedied or correct torqueing of the head bolts will not be achieved.

The first step toward fixing the bolt problem is to use some Moroso offset head-to-block dowel pins. These push the heads farther up the block face and give better manifold alignment later during assembly. They will also reduce or possibly eliminate the need to remove material from the long bolt holes. These offsets are available in 0.030- or 0.060-inch offsets. I have found that for the most part a 0.030 offset gets the job done.

To determine approximately what may need to come off the head faces, your first job will be to "check lap" the valves to see that they will, in fact, seal up. If the valves don't lap in virtually right away, your valve job has not been done properly, so take it back to the shop that did it and have it rectified.

With the seats OK for use grease up the stems and seats on the valve head and install the valves. Next set the heads level on the bench using a spirit level. From a graduated container, you can now pour windshield washer fluid into the chamber. Doing all this to establish what CR can be achieved and how much needs to be machined off may be something you can side step. An increasing number of machine shops are becoming performance orientated and this could make life simpler. All you need to do if you are dealing with such a shop is to take your heads in and tell them what you think is needed in the way of chamber volume and they will a) advise you if this can be done and b) check the volume at interim points during the machining to establish when the required cc has been achieved.

CR Increase Versus Cost

At this point it's time to tally potential costs of doing our CR business. Angle milling costs a lot more than flat milling. Not only is the setup time about 50-percent longer than for flat milling but we also must include setting up the head to machine the intake manifold face back to the correct angle with the head face. All this means the cost of angle milling can exceed the price of a 3.75 stroker crank which, with flat-top pistons and a 60-cc chamber will net about 11.5:1 or 12.5:1 with about 1/8-inch raised piston crown. What we are discussing here is relevant to any pair of heads that you may have stripped and modified, be they factory or aftermarket. After all the porting and machining has been done the heads can be readied for assembly.

At this point I need to introduce you to a CR solution using aftermarket heads. There is good news is this area–namely that they are available with chamber volumes as low as 49 cc. With a flat top in a 350 this would give about 12.3:1 compression.

CC'ing Equipment

Buying pro-style head cc'ing equipment can be a little expensive, and as just stated, may not be necessary unless you are contemplating doing the job repeatedly. With the ratios we'll be dealing with, most commonly 10 to 13:1, a graduated burette is nice but not totally necessary. Power House tools has an economy cc kit (part # 4975). This comprises a head plate and a tall graduated beaker. If you want to upgrade to a pro-style setup (part # 4974), you get a nice 100-cc burette, a stand to hold it, and the head plate.

To prepare the head for measurement, install the greased-up valves (grease seals them) and a spark plug, and set the head level end-to-end

This head is being leveled by means of a dial indicator prior to flat milling it 0.030 inch to achieve the desired CR.

using a spirit level. Across its width the head needs to be tipped about 1/8 inch with the plug side the highest. Now grease the plexi-glass head plate and place it over the chamber with the fill hole at the highest position so as to best allow air to escape while filling. Whatever the burette reads when the chamber is full is what you have. Add this to the cc contained in the block, piston cut-outs, and the head gasket and you are ready to calculate the CR.

To find out approximately what needs to come off, set the head level for flat milling or at about 1 degree (high on the plug side) if angle milling. Now fill the chamber with the amount of fluid to represent the volume needed for the CR required. At this point, you will need to measure how much the fluid is down from the head face. This will be a little problematic because when the end of the dial caliper approaches the fill fluid it will jump up to meet the end of the dial caliper. This means a little judgment must be exercised when making this measurement.

Cylinder Head Detailing and Assembly

What we are discussing here is relevant to any pair of heads that you may have stripped and modified, be they factory or aftermarket. After all the porting and machining has been done, the heads can be readied for assembly. Your first job, using a fine round or half-round needle file, is to remove all the sharp edges from around the combustion chambers. At the time the valves were check lapped they should have been numbered and placed in a 2 x 4-inch wood block suitably drilled and numbered to accept the valves.

At this point, the heads and everything going into them should be thoroughly cleaned. Spray every-

thing down with Gunk and go through the guides with a rifle-barrel brush. Next, brush everything else with stiff bristle brushes until all signs of dirt and debris are gone and then hose everything off.

Next, blow off the water with an air line. If you intend to paint the heads, now is the time to do it. Here I recommend using brake caliper paint as it stands up to the heat around the exhaust port better. Be sure you degrease the heads with lacquer thinner or the like before painting. Also mask the areas that need to stay as bare metal. After spray painting the heads place them in a warm environment (hot sun, warm oven,

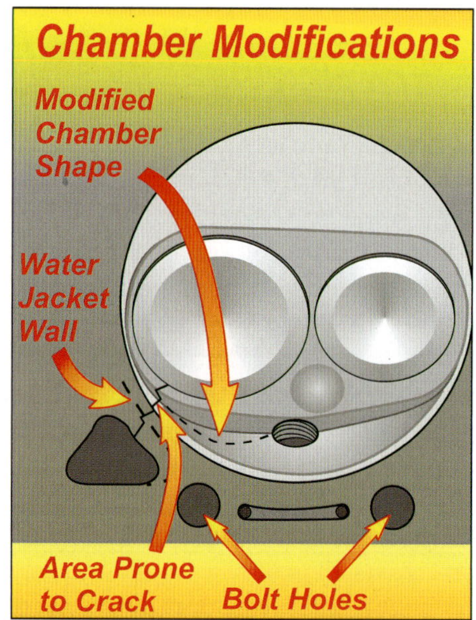

Fig 6-9. Apart from the basic de-shrouding of the intake and exhaust by sweeping out the chamber around the valve, the 186-style casting also benefits from some chamber reshaping. Be aware that these heads are prone to cracking at the point indicated here, so metal removed in this area other than sweeping out the chamber should be strictly limited. The area between the chamber and the bolt holes is solid and thick enough to be cut to the form shown.

etc) until the paint is good and hard then (and only then) WD-40 the machined surfaces to prevent rusting.

Assembly

Using a high-pressure lubricant, such as moly grease, smear the valve stems and the guide bores. Don't let there be any real excess on the intake valve as this grease will wash off and get on the spark plug. A little will burn off right away but a lot won't.

At this point, you can check the installed spring height that your combination of parts will give. Don't forget that the valve seats may be lower in the heads than stock, so the installed height of the spring will be greater. It is very important that the spring force produced when the valve is on the seat is as per the cam manufacturer's specifications. Fail here and the valvetrain can experience premature valve bounce. The stock spring height is typically 1.7 inches, but your choice of cam and the springs required to run it, may call for 1.750 or even 1.800 inches. In some cases where a bigger cam is used, the spring bases may need to be machined deeper (one more money-absorbing op).

Measuring the installed spring height is best done with a spring height micrometer, but a passably accurate job can be done with a dial caliper. Each spring station will require shimming to give the correct installed height. Generally, almost all heads will need something in the order of a 0.030-inch shim under the valve spring.

If the cam you are using calls for a dual spring installation, a spring locator or spring cup will be required to stabilize the base of the spring, so it does not walk around when in use. Spring locators/cups generally have a base thickness of 0.060 inch. If more shimming is needed to get the correct spring preload, install the shims under the spring locators/cups. For most of the applications, we are dealing with here getting the seat preload absolutely dead on is overkill. If the spring height is within minus 0.020 and plus 0.010 things will work out just fine.

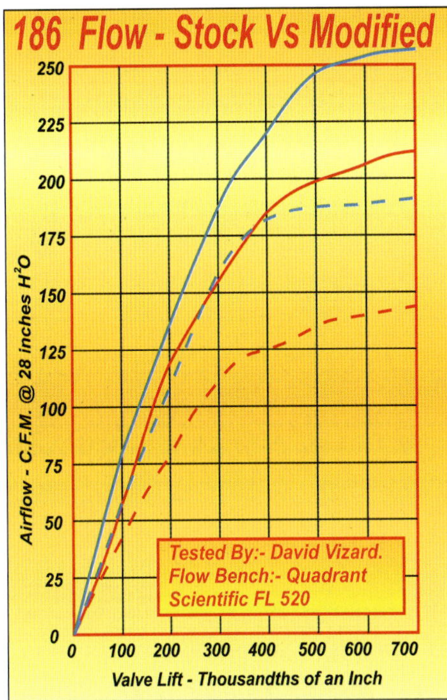

Fig 6-10. If you have the experience, equipment, and a willingness to take on a tedious job, here is what can be done with a set of 186 castings. The stock head (dotted) curves are blue for intake and red for exhaust. They show the flow with 1.94/1.5-inch valves. The solid curves are for a fully ported head using a 2.02/1.6 combination. These heads made 460 ft-lbs and almost 550 hp from a single 4-barrel 350 with a 12.5:1 CR and a 298 single-pattern race roller delivering 0.650 lift at the valves. These heads represented about a hundred hours of work. Later, better factory heads (i.e., Vortec castings) will produce much better results with only a quarter of the time to do the job, as will the latest budget aftermarket iron heads even in as-cast form.

However, the seemingly wide height tolerance just quoted does not mean we can skip the height check or ignore setting it right. Incidentally, if you are going to do the job like a pro, each spring needs to be checked and possibly detailed. Just how much will be needed will depend on the type and quality of the spring, so be sure to read how to prep your spring package in Chapter 7. The last point of possible problems is interference. Check that the bottom of the retainer does not run into the guide seal and also that at full valve lift the springs do not coil bind.

Once the heads have been assembled, bag them in a plastic bag and set them aside ready for installing on the rest of the engine.

Porting 186/041-Style Head Castings

The porting on these heads is straightforward enough given the guidelines set out elsewhere in this chapter. The longest job is carving out the excess material that resides in the bowls and around the base of the intake valve guide bosses. Be prepared for a lengthy haul even with a carbide.

Let's cover some of the pertinent points at the manifold face.

Don't be tempted to match the port at the manifold face with a large port gasket. It serves no useful purpose and can, as often as not, cut performance. I recommend using a Fel-Pro intake gasket part # 1204 as a template. As for the rest of the port, cut it using the guide lines given throughout this chapter. My experience is that, armed with the information given, most novice porters will cut a respectably good intake port of about 225 cfm.

As for valve sizes you can, if the cash is really short, stick to the 1.94 and 1.5 combination, but that is only practical if all the guides and the valve stems are good. If valves and a guide job are called for, then it's a 2.02/1.6 combo you should work toward.

Going the big valve route though is not without its problems as doing so increases the valve shrouding. So why don't we simply grind a chamber shape to minimize shrouding? Simple: the heads are more likely to crack!

Modifying the combustion chambers of the heads in question is one of the more critical areas if any kind of street reliability is to be had. These heads have a propensity for cracking between the intake seats and the water jacket in the area shown in Fig 6-9 especially if high compression ratios are to be used. Do not grind away too much metal in this area. Not only does it increase the possibility of cracking, it also has little positive effect on the airflow achieved. However, grinding 1/2 to 3/4 inch nearer the spark plug does increase flow, and that area is solid cast iron so you may as well go for it.

To keep the cracking situation to a minimum I recommend that the tool used to sweep out the chamber has at least a 1/8 corner radius.

Now is time to deal with the exhaust ports. To modify these use Fig 6-7 as a guide toward producing an effective exhaust port. Doing so should net about 170 cfm at 0.500 valve lift. Given a flow bench to reference the work and adding a high dose of experience can net a port that delivers as much as 195 cfm at 0.500 valve lift and as much as 210 at 0.700 lift. See Fig 6-10 for the results.

Porting L98 Corvette Heads

The L98 factory aluminum heads are, at 22 pounds, light and thin (iron heads are typically 47 pounds).

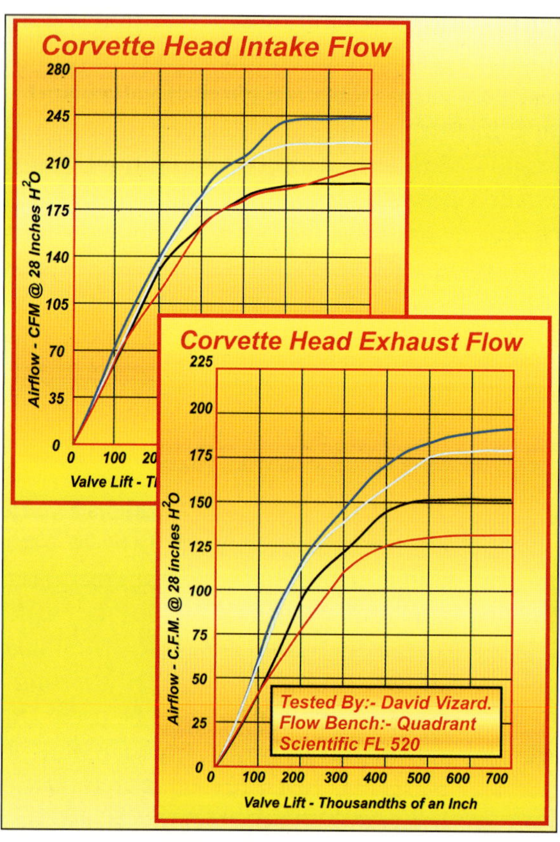

Fig 6-11. The red curves on these graphs are flow results for a stock 186 iron casting. The black curves are for an aluminum L98 casting. Note the intake curves of each of these heads, in stock form, are very similar. The exhaust of the L98 heads, though, is measurably better. Applying a basic porting job and retaining the stock valves delivers results of the light-blue curves. Installing bigger valves, re-cutting the seats to a higher flow design, and then reworking the entire port will deliver the results shown by the dark-blue curves. With a hydraulic flat-tappet cam of 260 degrees advertised duration, these heads delivered almost 400 hp and 455 ft-lbs from a two-plane-intake 10:1 engine. The torque curve was very strong right off idle and would have made a great truck motor.

the goal and peak RPM is not much more than say 6,000 for a 350 (and correspondingly higher or lower for other displacements) then these heads, when ported, may be a budget answer for your needs.

Because they are aluminum, these heads can be cut at about three times the rate to cast iron, so porting is much faster. Second, the reason they make good street heads but not good race heads is that the ports cannot be made too much larger without finding water. If you are looking for heads with a smaller chamber volume, these heads are typically around 58 cc (64, 68, and 76 are common chamber volumes) but milling is limited to a flat mill of about 0.020 because the deck face is relatively thin.

With the limitations just mentioned in mind, let's start with the combustion chambers. The first step here is to remove the casting gauging points between the spark plug and the head face. This together with a

Because of the effort to save weight, these heads are not as tough as their Bowtie brethren or any of the aftermarket heads. Also note that aluminum corrodes with age especially if the anti-freeze mix was not maintained properly. They can be very effective for specific uses if you can find a good set. However be aware they won't last any length of time over about 11:1 CR.

The flow of these heads in stock form is really no better than a typical early factory performance iron head. As a straight bolt-on (and with the appropriate self-aligning rocker), these heads have little if anything to offer. However, if you are looking for a cheap set of light heads for relatively easy porting, you might have something worthwhile. If the intent is to produce a true street combo where output from a little over idle is

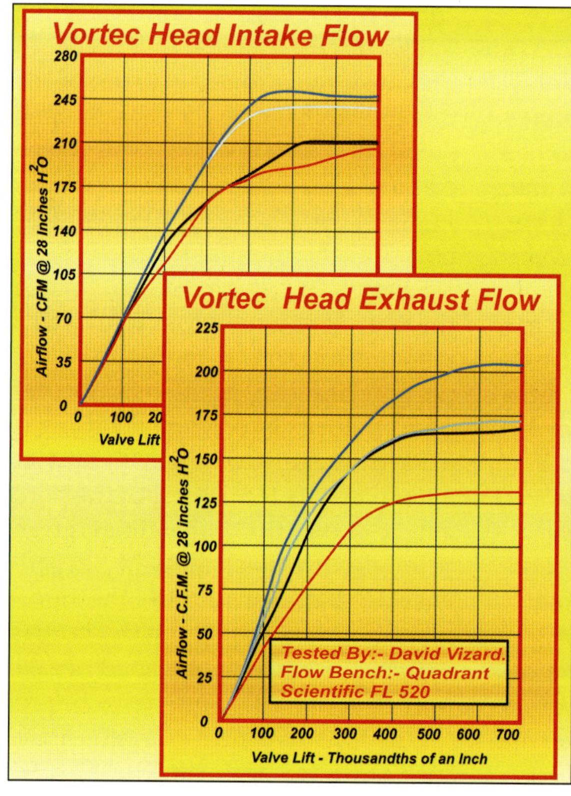

Fig 6-12. The red curves show the flow of the intake and exhaust of a stock 186 head casting. The black curves are for a stock Vortec head. The light blue curves are the result of installing 2.02/1.6 valves in place of the stock 1.94/1.5 valves and pocket porting the combination. The dark blue curves are for the ports after a basic but full rework.

chamber sweep around the valves (best done by your machinist) increases the chamber volume to about 60 cc. This volume together with a decked block 350 build using flat-top pistons will give a 10.8:1 CR. If you are building a 383 then for a 10.5:1 CR, a dished piston having about a 13-cc volume will be needed.

These heads normally have cast-iron guides and these guides more than likely need to be re-sized or replaced. This means you may as well go whole hog and install larger valves. The seats limit the maximum valve sizes to 2.00 for the intake and 1.55 for the exhaust. I have, in the past, cut down 2.02/1.6 valves for this, but some valve companies may well have off-the-shelf valves these sizes. I have flow tested with stock valves but never used any stock valve Corvette heads on an engine. With suitable valves on hand all the time, it never made sense not to go the 2.00/1.55 route, and that is what I would recommend here.

The L98 'Vette head's greatest asset is its high swirl and intake port velocity. To get optimal results, you should not detract from those assets. Though small, the stock port shape is not that bad for that era of production heads. Its principle shortcomings are casting flaws and small cross-sectional areas. The outside walls (see Fig 6-6) are the busiest areas, and consequently, are the place to work on. Using Fig 6-6 as a guide, work on wall "A" and its approach into bowl area "B" because this is the highest flow region and the most crowed as far as the air is concerned. Keeping the strongest flow on this side of the guide boss maintains this head's strong swirl characteristics. Next blend area "D" and just clean up the wall leading to area "C." Work area "C," so that it blends into the under-seat part of the port in a smooth fashion. At this point, rework the short-side turn into a radius as large as can be accommodated by the material present. These moves, with the bigger intake valve, should net the results shown in Fig 6-11.

The exhaust ports on these heads are relatively easy to do and deliver good results in the process. The point that must be kept in mind is that at mid- and high-lift the flow exiting the cylinder is from the center on out. This, in turn, means the port bias is important if good flow is to be realized at these higher lift values. Once this has been incorporated into your thinking, the rest gets easier. Just pay attention to the short-side turn and cut the rest of the port as per Fig 6-7 and all should work as intended.

Porting L31 Vortec Heads

In the absence of a flow bench, Vortec heads are about the most goof-proof castings to rework and get good results.

As seen from the flow charts in Fig 6-12, the stock L31 head is considerably better than the "Fuelie"-style heads we find ourselves working with if factory heads are the order-of-the-day. Because of the exceptional casting quality coupled with a highly effective design in terms of flow, porting these heads is about as simple as it can get. That's the positive side of things. The negative is the factory did such a terrific job in the first place, so there are only minimal gains to be had from pocket porting the intake.

The exhaust is a different matter. Because the intake flows well, the smaller deficiencies we see in the exhaust flow of the Vortec heads have a bigger negative impact on power output. For a production casting, the exhaust ports are pretty good, but they can be improved quite a bit.

It's all work for sure, but the good news is an exhaust port rework on a 375-hp motor can show an increase of up to 20 hp, and that's not too shabby considering you could get all the exhaust ports done in about 4 hours.

Because it is so easy to achieve professional-looking results, you may want to re-work all the ports. To do this, just take some 80-grit emery rolls and go over the entire port surfaces for a pro look. While doing so, pay attention to the pushrod pinch point as this can usefully be widened to bring the minimum port area up a little. This cleanup operation will start to pay off above about 0.400 valve lift. At about 0.600 lift, it can be as much as 5 to 7 cfm more.

The only way to see a significant increase from the intake in the 0- to 0.400-lift range is to replace the stock intake with a 2.02- or better yet a 2.05-intake valve.

The stock seat geometry is, like the rest of this heads design, very effective. Unless the seats for the bigger valve are done in an equally effective manner a lot of the extra potential of the bigger valves will be lost. Entrust your Vortec head seat work only to someone capable of producing an accurate and flow-effective seat.

Just installing the bigger intake valves does little to increase overall flow. The bottom line here is that if you are going to make the most of bigger valves the port needs to be reworked also. Again, if you re-work the intake port as per the guidelines of Fig 6-6, good results can be had. As for the exhaust target use the form shown in Fig 6-7, but leave the under-seat radius as stock. As for overall results, you can expect to achieve flow figures as per Fig 6-12.

Because the combustion chambers have a lot less shrouding than most production head castings, the

Vortec heads have better flow potential in the low- to mid-lift range, both on the intake and exhaust, but is partially penalized at lift values above about 0.500. This, and a relatively small intake port volume of about 176 to 178 cc, ported (170 cc stock) means good flow and port velocity. This, in turn, offers the potential to build a really wide ranging torque curve that produces strong HP everywhere, but it assumes the rest of the engine spec is compatible. A 10.5:1 350 peak torque number in the 445 to 460 range is possible along with power numbers in the 465- to 475-hp range.

A few words more on the subject of Vortec heads before we go on to aftermarket offerings. Please remember that the Vortec heads had a different intake manifold bolt pattern and require, in most instances, a Vortec-specific manifold. Also self-aligning rockers are used rather than guide plates, so be prepared to purchase another set of rockers if you don't have the appropriate ones.

Aftermarket Heads

When I wrote the first edition of this book, World Products made the only low-cost aftermarket heads, and these were cast iron. During the late 1990s, a shakeup among some of the leading lights in the cylinder head business started the snow ball rolling. Starting with Dart's Iron Eagles, we find that by about 2005 there were more aftermarket heads than you could shake a stick at. Some are very good, most are good and some are, well, let's say you won't see me using them any time soon.

Let's start with the World Products heads. At one time, I used a lot of these heads for my small-block Chevy builds for two reasons. First, they were the only game in town and

second, this meant I got a lot of practice porting them to fine tune what worked and what did not. By 2000 I could build a relatively cheap, simple, no-drama 600-hp road-race engine using these heads.

Basically World's heads fall into two categories: Stock Replacement (SR series) and the Sportsman series intended for high performance. The SR heads were originally intended to satisfy a market for heads to replace the later lightweight and thin highly crack-prone castings GM was producing from the mid to late 1980s on. The universal fit called for meant the original SR heads had small valves, but these heads were so well received that they felt justified in producing the SR Torquer, which was available in 1.94/1.5 or 2.20/1.6 valve-size form.

In unmodified form, they are okay but not intended for high performance, so one should not expect too much of them. That said, they port up surprisingly well, and if you come across a used set in good condition at a give-away price, they might be a real budget HP deal. I have ported a couple of sets of the bigger-

valve SR Torquers, and the final dyno numbers actually looked pretty good. Both sets were used on relatively low compression (9:1) truck motors with short cams (around 260 degrees advertised), and both sets made better than 430 ft-lbs on a 350.

I also built a Street Stock dirt-car motor that, in the cylinder head department, called for either stock factory iron heads up through to about 1985 or 1.94/1.5 SR Torquers. I opted for the Torquers here. The head chosen was the pivotal point of the build. With the highly constrictive rules, the front-running competitors were, according to the chief Tech man, finding about 335 hp max. After a very labor-intensive build, I produced a class-legal 368 hp. Now that may not sound like a lot of power, but the limitations imposed by the rules were hardly conducive to output. Things like a 0.45-valve-lift max. Absolutely no porting on the intake runners of the stock iron Q-Jet manifold. A stock Holley 650 or a stock Q-Jet carb, and the list goes on. Given enough time, I am sure I could have bumped that 363 hp number to 385 or maybe a tad more. Not that it

I built this SR Torquer-equipped, Street Stock dirt-car motor with a class-winning power output for about $2,500 and probably better than 250 hours of work. This shows that cubic effort is, to a large extent, a viable replacement for money.

would have made any difference as, with the hard spec tire called for, it would not, at any point around our local track, hook up anyway.

As far as World Products heads are concerned, my main push is with the Sportsman II heads. In as-cast form, they used to be the hot ticket for an easy bolt-together build that produced good results, but many later designs have overtaken them. If you find a set of used ones in good condition (they are almost indestructible), buy them. They are typically worth 40 hp over almost any stock casting other than the Vortecs. As for porting, they can deliver good results but because the casting is, what might these days be called "old school," the work is time-consuming due to the amount of metal that needs to be removed. Figure that pocket porting and slimming the guide bosses, which produces

good results, will take 4 to 6 hours. If you follow the porting rules given in this chapter, you can expect results as per Fig 6-13.

Aftermarket Performance Cast-Iron Heads

Since the first edition of this book I have concentrated on three brands of heads because the excellent results I am getting warrant doing so. With that said let's look at the options open to you among the heads that I have racked up flow bench and dyno time.

By employing what are essentially better and probably more expensive casting techniques Dart, Engine Quest (EQ), and RHS have produced heads that have sufficient casting-form and surface-finish accuracy to set new bench marks for heads with as-cast ports and chambers. Not only are the forms used far superior to

those available in the 1990s, but the surface finish is also far better. As a result, these heads have made life for the home porter so much easier that porting, between the old and new, is as different as night and day.

Out-of-the-box flow figures are as good as pro-ported heads from just a few years back. Because the design of the port and chamber form is so good these heads are far more sensitive to minor casting surface flaws. To better understand how this works, consider knocking a couple of dings into the intake port of an early-production casting. Because, it was already bad an extra unwanted imperfection made little difference to overall flow. Now imagine doing the same to the intake port of a $10,000 Cup Car head. What do you think that would do to the flow? If you suspected it could cost 20 cfm you'd be right. This also applies to our new-generation precision-cast heads. Because the ports are good in the first place, a lot more air goes through them and a seemingly minor flaw, when removed, can bring about a greater increase in flow than you can expect from a lesser casting.

For those who want to try their hand at porting heads, this is really a giant step toward the easy acquisition of pro-porting results. Just so you understand this point, let me be sure I have made myself totally clear: to achieve really effective results, you don't need a lot of fancy porting equipment, a flow bench, or a lot of experience. Instead, you need just average dexterity, a cheap die grinder, and some porting supplies.

Port and Chamber Volume Selection

Before you can start on any cylinder head project, you will need to make some decisions as to what port

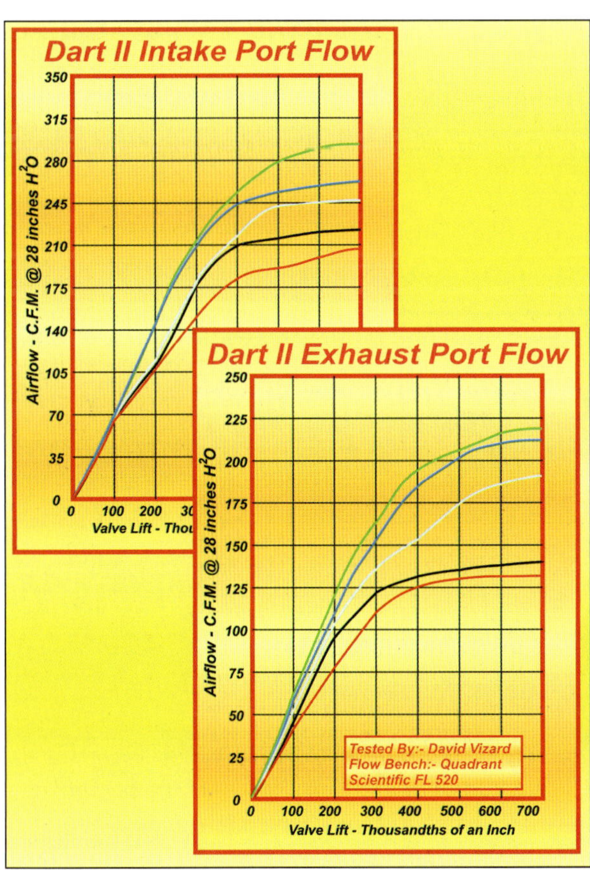

Fig 6-13. World Sportsman II heads are the original low-cost iron aftermarket heads. Although a little dated they can still make a good account of themselves. The red curves here are for stock 186 factory castings. The black curves are for as-cast Sportsman IIs. The light-blue curves are what you can expect by pocket porting and slimming the guides in these heads. The medium-blue curves are for ports that are to a basic full porting job. The green curves are what can be achieved with a serious re-work with a 2.05/1.6 valve combination.

and chamber volumes are likely to be best suited to your application. First, all of the iron heads we are looking at here will run a 10:1 CR on premium fuel without fear of detonation unless the engine is running too hot or the timing/mixture is off the mark. So, unless there is a good reason to use less compression, select a piston crown and a chamber volume combination that will give the required CR. As for the intake ports, please understand that bigger is not always better. We talk about port volumes here, but it's actually the port cross-sectional area that is the influential factor. The problem is that a small-block Chevy's intake port area varies quite a lot down its length from the intake manifold face to the valve. Since all 23-degree ports are fundamentally similar in length, this makes it easier to consider port size in terms of volume rather than cross-sectional area. To make a choice here use Fig 6-14.

Basic Porting Results

Here, I will describe the results seen by doing a basic porting job on EQ heads. However, the results seen with the Dart and RHS heads are, for all practical purposes, about the same. What this means is that the price is the deciding factor when choosing between these brands. You can be sure that some big parts house will have one or another of these heads on sale during any 12-month period. To reference all your choices in this area, check out the specs of the heads in the sidebar "Early Production Iron Performance Heads" on pages 63-64.

Having made your choice of castings, take them into your workspace and closely inspect them. It won't take long for you to appreciate the fact that in terms of shape 98 percent of the work is already done for you.

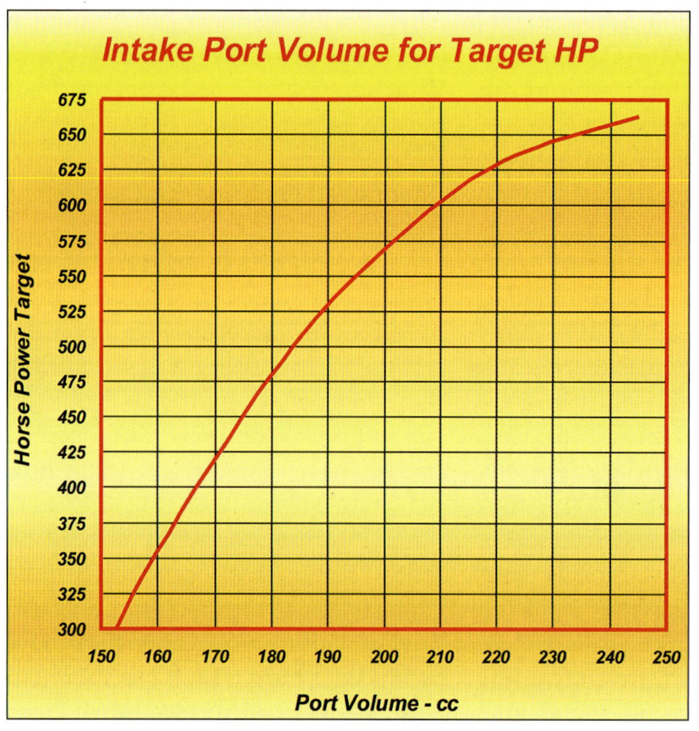

Use this chart to determine what intake port volume will be the best for your application.

In terms of finish quality, these heads are probably about halfway there before you ever start. To get results, the rules are simple. Just apply what was discussed earlier in this chapter. Since most of the work will be done before you start, you will find it is more like detailing rather than outright porting that you will be doing. Typical results can be seen in Fig 6-15.

As for output, a set of heads like this can deliver the airflow required for 540 hp for a flat-tappet 12:1 engine and as much as 585 hp for a roller-cam-equipped engine.

Iron Heads Compared

Here is my current experience with the three main players in the high-performance iron head business.

First, the RHS Pro Action heads. At the time of this writing, I am still building experience with these RHS

Here's a finished EQ 23 head. If you compare this with the one at the beginning of the chapter you will appreciate how little, in the way of metal removal, has been done to it.

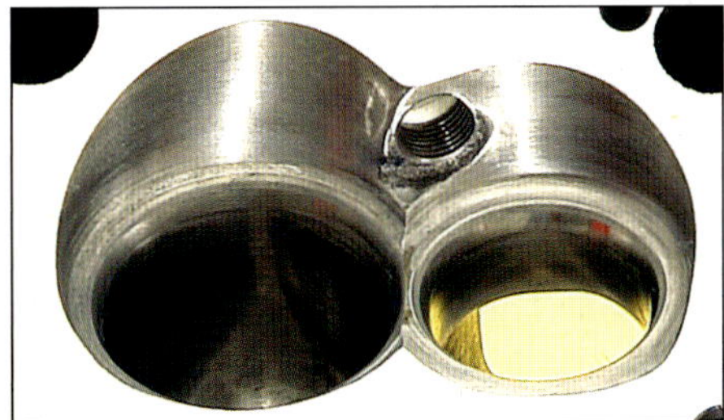

The re-working of the EQ 23 heads combustion chamber consisted mostly of detailing it as seen here.

heads and so far the results look really good. Out of the box, the mid-range flow on these heads is a little weaker than either of the other two we are considering here, but the top-end flow is a little stronger. Although this may be a minor factor in nature, it is something to take into account if the build is to have no more than, say, 0.500 valve lift. So far, all my experience with these heads has been with valve lift figures in the 0.600 to 0.625 range. Like the other two sets of heads under consideration here these heads, with basic procedures, port up very well.

DV Proven Power Heads

Cost-conscious circle-track race classes are the driving force behind the production of performance cast-iron heads. Both production numbers involved and the intensity of the competition has resulted in the development of genuine, low-cost, as-cast high-performance heads. To put their performance capability into perspective these heads, with little or no further porting, would have produced comparable results to those on a $40,000 1980 Cup Car engine.

Canfield, Edelbrock, and Dart have all, during my dyno testing, proven to be very strong performers when used with an appropriate combination of compression, cam, valve-train, and induction.

EQ 23 Flow Bench Tests

	#1		#2		#3		#4	
	Stock 180 cc		Stock 200 cc		Ported 2.02V		Ported 2.08V	
Lift	In	Ex	In	Ex	In	Ex	In	Ex
50	35	28	34	27	34	28	36	28
100	67	56	68	55	69	57	73	57
150	102	83	103	83	105	86	111	86
200	134	108	136	109	140	112	145	112
250	162	124	164	121	170	133	178	133
300	185	140	189	139	198	151	204	151
400	216	175	230	175	244	187	249	187
500	237	186	247	185	260	199	269	199
600	244	188	259	189	282	206	288	206
650	246	190	266	190	288	209	296	209
700	247	192	270	191	292	212	301	212

Tested By: David Vizard & Dusty Kennett

Fig 6-15. Test No. 1 was for an as-cast 180-cc port and 50-cc chamber head. Test No. 2 was as per No. 1, but with a 200 cc intake port. Test No. 3 was a basic rework of the ports, but with the original 2.02/1.6 valves retained. Test No. 4 was as per No. 3 but with the valveseat machined to accept a 2.08-intake valve.

RHS Pro Action Head Data
Test Head For Flow Figures – 200 cc Port Angled Plug

Lift	In.	Ex.					
0.05	35	27	Available Chamber Vols	50/64/72	50/64/72	50/64/72	50/64/72
0.1	68	62	Flat Mill Chamber Vols	45/58/66	45/58/66	45/58/66	45/58/66
0.15	103	85	Available Port Volume	180	200	220	235
0.2	133	101	Intake Valve Size	2.02	2.02	2.02	2.02/2.08
0.25	159	116	Exhaust Valve Size	1.6	1.6	1.6	1.6
0.3	187	133	Valve Lengths	Stk / +.10	Stk / +.10	Stk / +.10	Stk / +.10
0.35	211	151	Plug Orientation	Strt/Ang	Strt/Ang	Strt/Ang	Strt/Ang
0.4	231	163	Armored Exhaust Seats	Yes	Yes	Yes	Yes
0.5	264	100	Bronze Valve Guides	Yes	Yes	Yes	Yes
0.6	280	191	Screw in Rocker Studs	Yes	Yes	Yes	Yes
0.65	282	195	Valve Cover Bolt Pattern	Pre 87	Pre 87	Pre 87	Pre 87
0.7	281	198	Casting Finish	Excellent	Excellent	Excellent	Excellent

Dart Platinum Iron Eagle Head Data
Test Head For Flow Figures – 200 cc Port Angled Plug

Lift	In.	Ex.					
0.05	33.3	25.3	Available Chamber Vols	49/64/72	49/64/72	49/64/72	49/64/72
0.1	67	57	Flat Mill Chamber Vols	44/58/66	44/58/66	44/58/66	44/58/66
0.15	103	87	Available Port Volume	180	200	215	230
0.2	135	99	Intake Valve Size	2.02	2.05	2.05	2.08
0.25	163	115	Exhaust Valve Size	1.6	1.6	1.6	1.6
0.3	190	129	Valve Lengths	Stk / +.10	Stk / +.10	Stk / +.10	Stk / +.10
0.35	212	143	Plug Orientation	Strt/Ang	Strt/Ang	Strt/Ang	Strt/Ang
0.4	233	155	Armored Exhaust Seats	Yes	Yes	Yes	Yes
0.5	262	180	Bronze Valve Guides	Yes	Yes	Yes	Yes
0.6	272	191	Screw in Rocker Studs	Yes	Yes	Yes	Yes
0.65	274	197	Valve Cover Bolt Pattern	Pre 87	Pre 87	Pre 87	Pre 87
0.7	277	200	Casting Finish	Excellent	Excellent	Excellent	Excellent

EQ Cylinder Head Data Test Head For Flow Figures – 200 cc Port Angled Plug							
Lift	In.	Ex.					
0.05	33.5	25.6	Available Chamber Vols	50/64/72	50/64/72	750/64/72	50/64/72
0.1	68	64	Flat Mill Chamber Vols	45/59/67	45/59/67	45/59/67	45/59/67
0.15	103	81	Available Port Volume	180	200	215	230
0.2	136	99	Intake Valve Size	2.02	2.02	2.02	2.08
0.25	164	113	Exhaust Valve Size	1.6	1.6	1.6	1.6
0.3	190	129	Valve Lengths	Stock	Stock	Stock	Stock
0.35	211	137	Plug Orientation	Strt/Ang	Strt/Ang	Strt/Ang	Strt/Ang
0.4	230	147	Armored Exhaust Seats	Yes	Yes	Yes	Yes
0.5	249	173	Bronze Valve Guides	Yes	Yes	Yes	Yes
0.6	263	182	Screw in Rocker Studs	Yes	Yes	Yes	Yes
0.65	270	187	Valve Cover Bolt Pattern	Universal	Universal	Universal	Universal
0.7	274	191	Casting Finish	Good	Good	Good	Good

Dart's Platinum Iron Eagle is next. These heads benefited greatly from the use of an intense "wet-flow" development program. My dyno tests have shown this wet flow technology to be worth over 20 hp on a nominally 480 hp test engine. These heads, either as-cast or with basic porting, have shown some fine results. Like the other two head designs featured here, these heads port up very easy. Just before going to press with this manuscript, I ported a set of 230 cc runner versions of these heads (2.08 valve). The finished result was 315 cfm for the intake and 223 cfm for the exhaust at 0.700 (seven hundred thousandths) lift. To put those figures into prospective given a mild race cam, 12.5:1 compression and good race single plane induction system that's more than enough air flow to top the 600 hp mark from a 383.

Of the group here the heads I have the most experience with are the Engine Quest EQ 23 heads. For around $500, bare per pair, these heads are a great deal. The casting finish is definitely modern era but just shy of being as good as the either Dart or RHS offerings. But finish is far less important than form. From my dyno testing experience with these heads in both the 180- and 200-cc versions, I have concluded they have what it takes in terms of swirl, combustion efficiency, and wet-flow characteristics. As an interesting aside, there have been several highly regarded professional engine builders who have lost bets by underestimating the power production capability of these heads.

As well as being good out of the box, these heads port up easy (with 2.08/1.6 valves 301 cfm on the intake and 207 on the exhaust at just 0.600 lift) and can be done by a rank beginner over a weekend. Since their introduction in 2006, these heads have been showing championship winning form on many circle tracks.

I suspect that many of the iron heads discussed here will get a basic porting job done on them, bringing up the question as to which ones port up the best. Having gone down that road, I can tell you that after porting the results are so close that it's a virtual three-way tie in that department. If you are looking to port one or another of these head sets, then price at the time you intend to purchase should be the deciding factor.

Aluminum Heads

Without a doubt, there are some really good heads that have become available since 2000. I have run quite a few pairs of Canfield heads, a couple of pairs of Edelbrock heads, and a ton of Dart heads in as-cast and subsequently ported form. In addition to this, I have run CNC-ported Dart and AFR heads with really good results. However, only one model of CNC-ported head falls into the price range set for this book: AFR's 195 Eliminator street head.

All of the heads I have used over the past ten years have been the result of recommendations by reputable engine builders who have had successes with the heads concerned and have recommended I try them. This makes life easier for me as magazines are never in a rush to run a story on a product that does not work. What they really want is something that produces results good enough to get excited about. Well the advice I got from my pro engine builder friends saved me for the most part, testing performance equipment that might have otherwise been indifferent. Let's start with the Canfield heads.

Canfield Heads

For a 350 or a 383, I have favored the 195-cc version of the two port volumes offered for Canfield heads. They have produced good results in out-of-the-box form and responded well to basic porting. With a single-pattern flat-tappet Comp Cams cam (276-3 profile on a 108-lobe centerline angle) 430 hp and 440 ft-lbs is on from a street-driving 10.5:1 350. With a hydraulic roller cammed 383, this goes up to 470 ft-lbs and 470 hp. From engines of a simple bolt-it-together spec, these are good numbers. But these heads can be basic ported over a weekend and, in round numbers, that's typically good for 25 to 30 hp. When ported, these heads will top the 515-hp mark along with about 485 to 490 ft-lbs of torque without a huge expenditure of cash. A Comp Xtreme 224/230 at 0.050 on a 106-lobe centerline angle and 1.6/1.5 rockers does the job in the valvetrain department for this kind of output. Couple that to a Super Victor intake and a good 800-cfm carb, and you are in business. Also noteworthy is that I have seen very good results on a 350 equipped with an Edelbrock Air Gap Performer intake.

Canfield also makes a 220-cc version of their small-block Chevy head. These are good as-cast for racier 383s and bigger engines.

Ported they can really deliver results. At 12.5:1, a solid roller 383 can be coaxed over the 600-hp mark without too much drama. Although a little outside the scope of this book, I did do a set of these to an all-out spec that went on a not-so-budget 13/1 440 inch small-block that went well past the 700-hp mark.

Edelbrock Heads

Edelbrock offer a wide range of options on their street heads, and I figured it would be a full-time job for quite a while testing them all. It might be nice to do just that, but the time and budget just aren't there. My main test objectives with the Edelbrock Chevy heads I have used are for street performance where true low-speed street output is called for. Both the emission-legal Performer and the Performer RPM heads have produced power curves on either a 350 or a 383, which did just that. As they come out of the box their 170-cc intake runners produce strong dyno numbers in the 1,500- to 3,500-rpm range but tend to fall off much above that. Spend a weekend porting them, and you will like the results. Doing so makes the ports only about 3 to 4 cc more, and you will like what it does to the top-end output. Essentially, a 383 will pump out about 30 hp more with no measurable loss at the low end.

If you are building a 350, it's worth noting that these heads don't come in the smaller chamber sizes. You will be looking at 70-cc chambers for the Performer and 72 cc for the RPM version of such. This means achieving a 10:1 CR is more of a problem unless the build involves more than 350 inches.

Dart Heads

I have lost count of the number of Dart aluminum Pro 1 small-block Chevy heads I have used on various mule/project engines. With an appropriate parts combo, they worked well as-cast and especially well after I ported them. Since I am in no rush to fix something that works, I have used ported Dart heads on a lot of my engines because I had done enough of these heads to know where to easily increase flow and, consequently, HP.

In 2006 Dart came out with their wet-flow-developed version of the Pro 1–the Platinum Pro 1. With these later heads, Dart was talking an extra 15 to 25 hp depending on the combination it's run on. Sounds good, but I needed my own numbers just to establish the value of whatever wet-flow development they had done.

Canfield 195 Head Airflow			
Lift	As-Cast		Basic Ported
	In.	Ex.	In.	Ex.
0.05	35	29	35	28
0.10	71	59	72	59
0.15	102	81	104	83
0.20	139	102	144	108
0.25	168	122	174	130
0.30	193	139	203	146
0.40	242	170	248	178
0.50	256	185	263	195
0.55	259	188	270	203
0.60	257	189	279	206
0.65	256	189	282	209

Canfield 220 Head Airflow			
Lift	As-Cast		Basic Ported
	In.	Ex.	In.	Ex.
0.05	36	29	35	28
0.10	71	60	73	59
0.15	104	81	107	82
0.20	144	102	148	109
0.25	173	124	177	131
0.30	204	140	209	147
0.40	248	170	255	178
0.50	260	186	272	197
0.55	265	189	279	204
0.60	270	190	288	207
0.65	271	191	293	211

Edelbrock 170 cc Port Performer Airflow			
Lift	As-Cast		Basic Ported
	In.	Ex.	In.	Ex.
0.05	34	26	35	27
0.10	64	56	67	59
0.15	100	73	103	76
0.20	131	99	141	104
0.25	156	115	165	122
0.30	179	133	190	144
0.40	220	162	239	173
0.50	237	180	245	192
0.55	240	182	252	199
0.60	232	182	254	202
0.65	234	180	246	202

Edelbrock 170 cc Port Performer RPM Airflow			
Lift	As-Cast		Basic Ported
	In.	Ex.	In.	Ex.
0.05	33	27	34	27
0.10	63	54	67	58
0.15	100	75	104	77
0.20	134	99	142	104
0.25	158	115	166	121
0.30	182	131	192	143
0.40	224	163	241	174
0.50	241	179	249	191
0.55	244	180	255	200
0.60	230	181	254	204
0.65	230	182	246	205

The interesting aspect of this test is that the flow of the 200-cc Pro 1, and the later Platinum versions were virtually identical. In this instance, any differences in output would be due solely to the hopefully better wet-flow characteristics of the Platinum heads over the originals.

In this instance, the test engine was a 383 with a hydraulic roller valvetrain. Fig 6-16 shows the results. You can see that some applied wet-flow technology worked very much in our favor.

Although I have used all the port runner sizes of Pro 1 heads that Dart offers, my main experience is with the 200 and 215 heads. But before I discuss that let's talk about their 180-cc heads first. If you are building a street 350 or a 383 that has to have low-speed grunt, these heads work fine right out of the box. They have 500-hp capability but work out best if cammed to give about 480 hp. With a cleanup 500 hp is almost a breeze and can be achieved without compromising low-speed output especially if the build is a 383.

While the 180-cc Darts work very well out of the box or in basic ported form I do have a move that makes for a wider power band and a little more upstairs without a low-speed sacrifice while still delivering a good top end. This works well if you are building a 350 or a 383 that really must be a nice street driver.

At the top is the chamber and intake bowl of the original Dart Pro 1 heads. Below is the wet-flow-developed Platinum Pro 1.

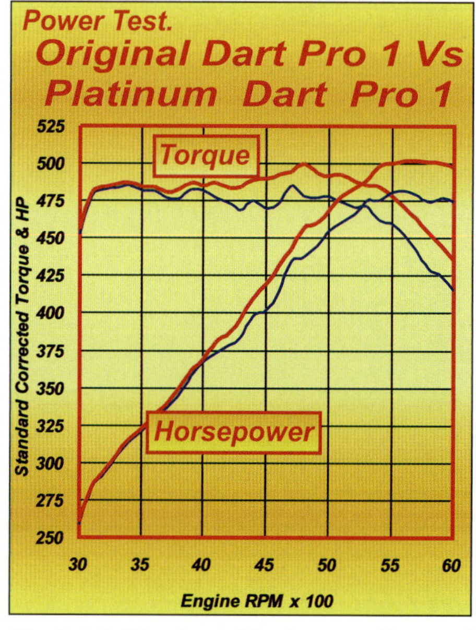

Fig 6-16. The benefit of correcting wet flow problems as much as possible is clearly shown by these results. Dart's Platinum head (red curve) benefited to the tune of 20-plus hp.

Dart Platinum Pro1 180 cc Port Airflow				
Lift	As-Cast		Basic Ported	
	In.	Ex.	In.	Ex.
0.05	32	27	34	27
0.10	61	59	66	62
0.15	96	78	103	87
0.20	131	100	144	110
0.25	159	114	166	139
0.30	180	130	194	161
0.40	222	162	240	188
0.50	244	182	253	200
0.55	246	184	263	206
0.60	248	188	269	208
0.65	243	190	272	209

Dart Platinum Pro1 230 cc Port Airflow				
Lift	As-Cast		Basic Ported	
	In.	Ex.	In.	Ex.
0.05	34	27	39	27
0.10	69	62	77	62
0.15	109	89	113	88
0.20	140	108	148	112
0.25	168	122	178	142
0.30	192	141	204	167
0.40	238	175	251	196
0.50	275	193	289	211
0.55	291	199	303	215
0.60	293	200	310	219
0.65	294	201	315	222
0.70	294	202	318	224

Introduced late 2006, the Dart Platinum head benefited from an intense wet-flow development program.

Basically, I replace the 2.02 intake valve with a 2.08 valve, cut a high-flow seat and then apply a basic porting job. Using this head, I then spread the cam's lobe centerline angle by 1 degree from what would have otherwise been optimal then pick profiles for the intake and exhaust about 4 degrees shorter. All this results in the same top end as the smaller valve and longer cam, but with a smoother idle, more torque in the low speed range—and better mileage!

If you are building an engine where you are targeting about 440 to 450 hp, these heads need to be a serious consideration. With basic porting, they are good to 500 very streetable hp on a 383. The top chart on page 84 shows the flow numbers you can expect.

As for large runner heads, I have ported some 230-cc Darts for some serious HP on big-inch small-blocks. Even with streetable cam specs and 10.5:1 CR, these engines top 600 hp with ease and deliver torque numbers of 575 ft-lbs from 440 inches. The bottom chart on page 84 shows the flow specs for a 230-cc head, stock and ported. From these figures, you can see that either out of the box or

Note these ported 230 Darts are not big on shine. Surfaces were finished with 100-grit rolls. Flow was at 0.700 lift, 318-cfm intake and 224-cfm exhaust. This came with strong low- and mid-lift numbers making them good for a big-inch engine.

in ported form the Dart Platinum 230 acquits itself well.

The most likely size of engine that a reader of this book will build is a 383, and that is very much the same for me. That is why I have had so much more experience using heads of 200- to 215-cc intake runner volume. Again, assuming street usage I tend to favor a 200-cc port if I am going to port the heads. If I am not going to port the heads, I tend to favor the 215s. So that you can see how, in as-cast form, each size of port volume Dart head responds on a typical bud-

get street driver check out the results in the sidebar "Port Sizes" on page 86.

OK. Let's take a look at the flow figures for the Platinum versions of these heads. The 200-cc head responded to our flow bench as shown below left. Using a two-plane intake manifold and a relatively budget valvetrain, such as a flat-tappet hydraulic cam of, say, 270 degrees seat-to-seat duration, you can expect a 10.5:1 383 to deliver 485 ft-lbs and about 435 to 440 hp using these heads as-cast. Port them and 495 ft-lbs and about 475 to 485 hp are yours to be had.

Dart Platinum Pro1 200 cc Port Airflow				
Lift	As-Cast		Basic Ported	
	In.	Ex.	In.	Ex.
0.05	32	27	34	27
0.10	61	59	65	62
0.15	96	77	102	87
0.20	129	101	145	110
0.25	159	114	168	139
0.30	179	129	190	161
0.40	220	164	240	187
0.50	249	185	257	201
0.55	256	190	271	205
0.60	260	193	275	207
0.65	261	195	277	209

Dart Platinum Pro1 215 cc Port Airflow				
Lift	As-Cast		Basic Ported	
	In.	Ex.	In.	Ex.
0.05	33	28	34	27
0.10	64	58	66	62
0.15	99	78	104	87
0.20	131	100	146	110
0.25	167	113	173	139
0.30	179	130	192	161
0.40	220	165	242	187
0.50	253	184	260	202
0.55	263	188	274	206
0.60	274	192	281	207
0.65	276	193	286	208

Port Sizes

Quoting port sizes by volume appears to have become the norm for two reasons. If we go back to the late 1960s and early 1970s, the only choice for heads was factory casting with ports for street applications. Having less-than-ideal castings to work with meant a lot of grinding to maximize the castings capability. To give the customer an idea of how much work had been done, the head porters would quote port volumes as a measure of what had to be taken out to achieve the end product. Bigger equated to more work and thus the ability to justify a higher price tag.

Since these early castings were limited, the more metal I removed from the ports, the faster the car went, and the sooner the customer would be back for more ported heads to replace those that cracked. And believe me, cracking had to be an accepted outcome if you wanted to go fast.

So demonstrating the amount of work done on castings was one reason for quoting port volume. The other, and initially secondary factor, was to get some idea of the size of the cross-sectional area of the port.

Since the port area changes substantially as the port progresses from the manifold face to the valve, quoting size in square inches is not practical. That left port volume as a viable option. The bigger the volume, the bigger the mean cross-sectional area of the port.

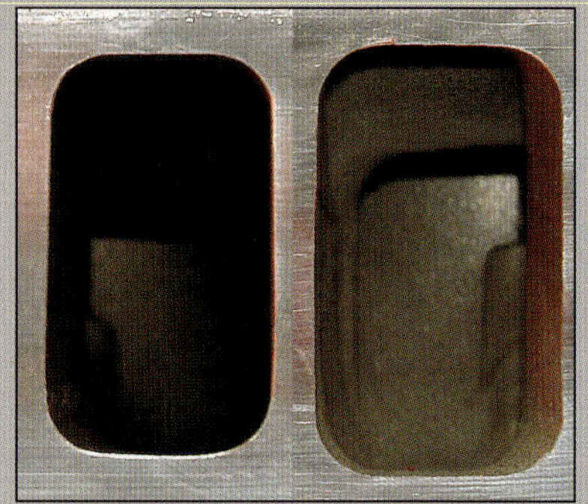

Shown here are the sizes at the manifold face for a 180-cc port runner (left) and a 230-cc port runner (right).

So why is port cross-sectional area important? The flow surely goes up if the area is bigger, and that's what we want, is it not? Sure, the engine wants as much airflow as possible, but much of the flow through depends on port velocity and the generation of pressure pulses. This means an overly large port can hurt power even though it may, on the flow bench at least, flow better.

As can be seen from the nearby flow tests, the bigger port does flow more up at the higher valve lift numbers, but not necessarily so at the lower left.

Part of the increased high lift flow is due to a slightly bigger intake valve in the case of the 215- and 230-cc ports, but about 70 percent of the additional flow above the 0.500 valve lift point is due to the bigger port, not the bigger valve. So the big-port/big-valve combo flows the biggest numbers. The question is how does this work out on the dyno?

Characteristic Port Area

Here is what a size comparison of the characteristic port area's look like for our test heads. Increasing the area may increase the flow but the port velocity drops. This may hurt output more than the flow increases it.

Shown here is a comparison of the mean port size of each of the test heads. Black is for the 180-cc port, red 200, green 215, and dark blue 230.

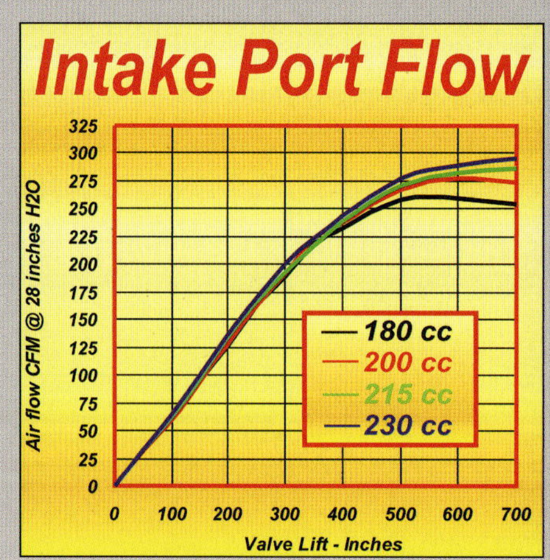

For all practical purposes, there is no great flow difference between these heads until about the 0.150-lift mark. At this point the superiority of the bigger ports starts to pay off. Even so, those differences are hardly significant until about 0.300 lift. To tap into the full potential of the bigger ports, a valve lift of at least 0.600 was necessary.

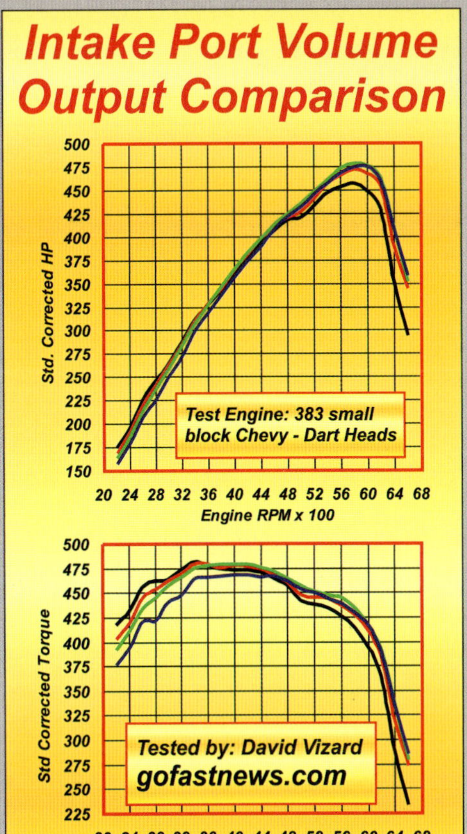

The results shown here clearly demonstrate that smaller and, consequently, higher-velocity ports, favor low-speed output. These results also show that going too big (blue curve of 230 cc port) on the ports for the job at hand produces worse results almost everywhere in the RPM range.

To better understand what is going on here it is best to view low-speed trends from the torque curves. To see what is working best at the top end of the RPM range study the horsepower curves.

Not only do these curves show a trend but they also demonstrate that choosing a head with too much port volume, as in the 230-cc example, delivers less power potential everywhere in the RPM range. From these results we see that the 215-cc port (green curve) equaled or beat the 230-cc port (blue curve) everywhere, thus proving bigger is not always better. Combining what we see from the torque curves and the HP curves, the 200-cc runner (red curve) proves to give the best average numbers over the RPM range tested for the particular spec that the rest of the engine had.

Dyno Time

Take a look at the graphs showing the 383 test engine's output. To more clearly appreciate the trends involved, the torque and horsepower graphs have been separated. The effect any particular head has on low speed output can be more clearly seen by considering the curves shown on the low end of the torque graph. To see what happens at the top end, look at the high-speed results on the HP graph.

As the torque curves show the 180-cc ports (black curve) produced the best output up to 3,400 rpm, peaking at a stout 482 ft-lbs. The 200-cc port (red curve) was not lagging by much and from 3,400 rpm up it ran up with (or close to) the bigger ports. If we look at the torque curves and also consider the horsepower curves we can see that the 200-cc ports delivered the best curve for our 383-incher (with the cam it had and the RPM range it would operate in). The 215-cc (green curves) heads made the highest horsepower by

pumping out 478 hp, as opposed to 457 for the 180-cc runners, 472 for the 200, and 475 for the 230s. The price the 215s pay over the 200 to achieve this 7-hp advantage is that they give away up to 10 ft-lbs of torque from 2,300 to 3,200 rpm.

The 230-cc port runner heads failed to deliver any worthwhile superiority anywhere in the RPM range on this particular test engine. The smaller 215-cc port heads beat the 230s everywhere! This shows that an engine is not quite the simple air pump it is often touted to be, and that bigger is not necessarily always better. The bigger port heads would have paid off had we targeted an engine capable of more RPM or one with bigger displacement. Experience with ports in the 230- to 245-range show that every bit of the port size is needed if you are building a 700-hp 440-ci small-block. If we look at a comparison on a pro-rata basis a 235-cc port on a 440-inch small-block Chevy is only equivalent to a 186-cc port on a 350.

So how do you decide what port volume your small-block Chevy should have for best results? Answer: Use Fig 6-14 shown earlier in this chapter.

Just remember that too small a port will be a far better deal to drive than one that is a little too large. A port that is 20-cc too big can easily cost 20 ft-lbs at a point in the RPM band that is most often used for a true street driver.

Just in case you are wondering, with a bigger cam, 10.5:1 CR, and our 200-cc Platinum Darts, this test engine cranked out right on 500 ft-lbs and a tad over 502 hp.

Stepping up to the as-cast 215s for a 383 and using a hydraulic roller of about 240 degrees at 0.050, a thoughtfully built 383 can return some 500 ft-lbs and power in the region of 520 to 530 hp. My best effort to date was a 244 at 0.050 Comp single-pattern hydraulic cam 383 with as-cast 215 heads that made 495 ft-lbs and 528 hp. Ported they will go about 25 more horses than this. The flow for both as-cast and basic ported Dart 215 Platinums is at bottom left on page 85.

AFR 195 Eliminator

The Eliminator from Air Flow Research is our entry-level CNC-ported cylinder head. At the time of this writing, I have had experience using this head on 355- to 408-ci engines, and it has produced very satisfying results each time. Introduced in its current form about 2005, this head design has established itself as one of the most successful on the market. In addition to this, it is also one of the, if not the most, cost effective of CNC heads available. In fact, at the time of this writing, it is the only one that comes in at a price that allows it to be used within our budget constraints.

I have used these heads on a 383 build and closely approached the 600-hp mark, yet still had change left out of a $5,500 bank roll. These heads come with 8-mm-stem lightweight valves, smaller-diameter valve springs of much higher quality than you would expect of heads in this price range, plus lightweight retainers. All these valvetrain factors make these heads very hydraulic roller friendly, and in simple terms, that means more RPM on less spring.

For more information on all the heads covered in this chapter go to MotorTecmagazine.net.

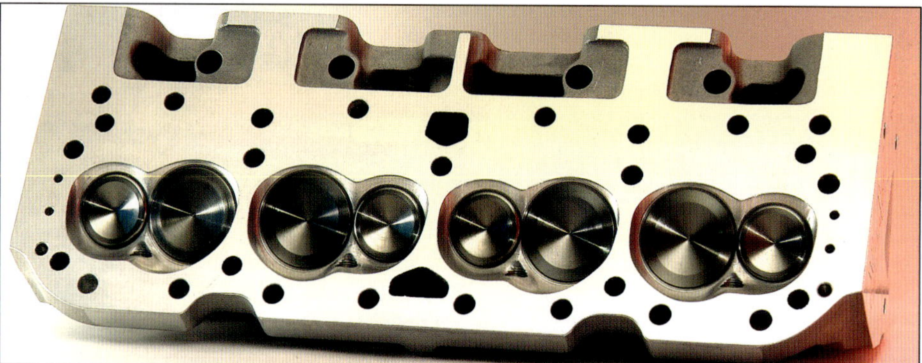

In addition to CNC porting, the 195 Eliminator also comes equipped with lightweight valves and special low-mass springs that deliver much better valvetrain control with a hydraulic roller cam.

Seen here are the ports and chambers of the AFR 195 Eliminator head. To keep production costs to a minimum a coarse feed is used on all surfaces other than the combustion chamber.

These exhaust and intake cross-sections show the AFR 195 Eliminator port forms. Note the large short side turn on the exhaust. This contributes greatly to delayed separation of the exhaust flow from the port floor under real-world running conditions. In turn, this cuts low-speed exhaust reversion and helps boost low-speed torque with no down side at high RPM. On the intake, the minimal "ski-jump" form on the port floor can be seen. The ski-jump floor looks better on paper than it does in the real world. The flatter approach to the short side, as seen here, often seems to help reduce wet-flow impact on the chamber wall adjacent to the spark plug.

CAMS & VALVETRAINS

A problem confronting many engine builders is making the best choice of cam and valvetrain components. Picking up a cam catalog doesn't necessarily help because of the number of choices. On a single page you may see several cam specs given, and many may seem fine if you base your choice on the application description. Maybe one has a little more lift than another, but a little less duration. Another may have slightly less duration, but be on a tighter lobe centerline angle than yet another possibility. Sure, you can ask people what's worked in their motors. Although that may seem a simple solution, let me assure you it can easily fail to produce within 30 hp of optimal results in 95 cases out of 100, and I'm talking about a 350 here. If the engine is bigger, the discrepancy can be even greater.

Why so far off? Ask yourself if the person you are seeking advice from has successfully built a motor like you intend to build. If so, do you intend to closely follow the specifications that were used? How do you know this person was qualified to give advice?

For the record, I've found that even among successful, professional, race-winning engine builders, real camshaft expertise is rare. I hate to sound like I am

The cam and the valve train components chosen are critical to the engine's success as a power producer. Stick to what you read in this chapter and your engine will be a success in this department.

blowing my own trumpet here, but this is a subject I teach at University level. Along with that, I have also redesigned a whole line of cams for several cam companies to very good effect. In two instances this was done to the extent that the engines in question made more power from my hot street cams than they had previously seen from their race cams! The advice you are getting here is based on 40 years of testing cams in general plus well over 10,000 combinations (8,000 in one six-month-long session alone for Crane) for small-block Chevys.

My recommendations are based on obtaining the maximum output, especially in terms of torque, for a given intake duration.

Basic Five

There are five parameters to be addressed when selecting a cam and valvetrain for best performance. In order of importance they are: overlap, the lobe centerline angle (LCA), advance/retard setting, lift, and duration. Understanding how each is affected by engine spec is the

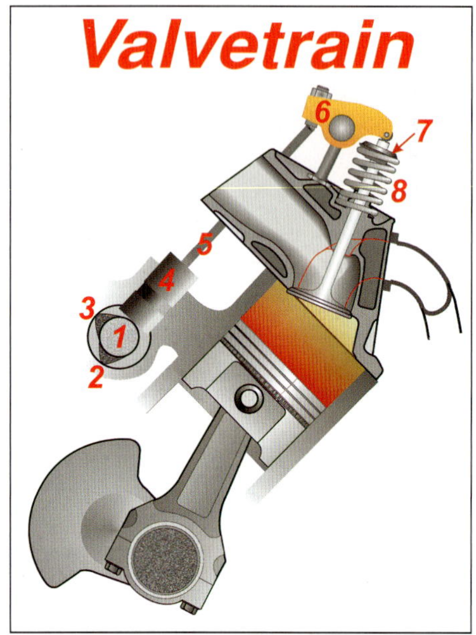

The eight factors dealt with in this chapter are: 1) Cam base circle diameter; 2) Exhaust lobe; 3) Intake lobe; 4) Lifter or tappet; 5) Pushrod; 6) Rocker; 7) Retainer-keeper-lashcap; and 8) Spring.

The aftermarket may have paved the way but the advent of the roller cam in stock engines has opened a new avenue for cheap power.

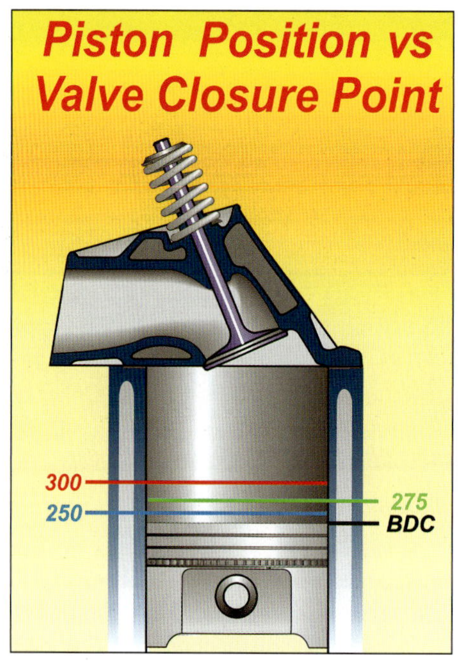

Fig 7-1. Here we can see from the piston position at different closure points, as dictated by ever-longer cams, how the dynamic compression drops as the cam gets longer.

key to getting the best cam for the job. Some engine configurations respond better to added lift than to increased duration, and some the reverse. Some engines need a tighter LCA than others, and so on.

Investigation reveals that cam characteristics are tied to cylinder head airflow capacity and the size of cylinder it has to supply. This implies that a flow bench is not only a tool for developing heads, but also for selecting a compatible cam design. To see how the factors of flow and displacement affect the cam spec required, let's start with the seemingly disconnected issue of fuel octane.

Fuel Octane and Dynamic CR

Fuel octane influences the compression ratio that can be used and an engine's CR must, for best results, be tied to cam duration. The key factor is the dynamic CR based on the intake valve closing point. This is closely related to the crank-

ing pressure seen on a compression gauge. Obviously the piston will be well up the bore if the valve closes late as per a long-duration cam. At low RPM at least, the cylinder will trap less air and therefore produce a lower dynamic CR. This causes reduced cranking pressure and considerably less torque, especially at the low end.

Fig 7-1 illustrates just how this lower cranking/low speed compression comes

about. We need the CR to be as high as possible without incurring detonation. Detonation is brought about by a combination of time, heat, and pressure. For most engines, detonation occurs at low RPM because of the greater length of time the end gases are exposed to the advancing flame front. When a long cam is used, the compression pressures at low RPM are reduced, thus not only allowing—but requiring—a higher CR to be used.

The static CR must be based not only on fuel octane but also intake valve closure point. This means establishing the dynamic CR required. For the record, most successful race engines run dynamic CRs of about 9:1, and a good street motor on service station fuel about 7.5:1. If the compression ratio isn't in line with the cam duration, the closure point of the intake valve, the dynamic CR, and cranking pressure, will consequently be low. This not only kills torque everywhere in the rev range, but also significantly reduces the engine's low-end output. In other words, it emphasizes the undesirable long-cam features that we're trying to avoid.

By putting more compression into the engine, it's possible to regain most of the low-end output lost by the use of more duration, and all of it in the case of shorter performance cams. A good test to establish that an engine has a high-enough CR is to do a cranking test. Assuming that rings and valves seal perfectly,

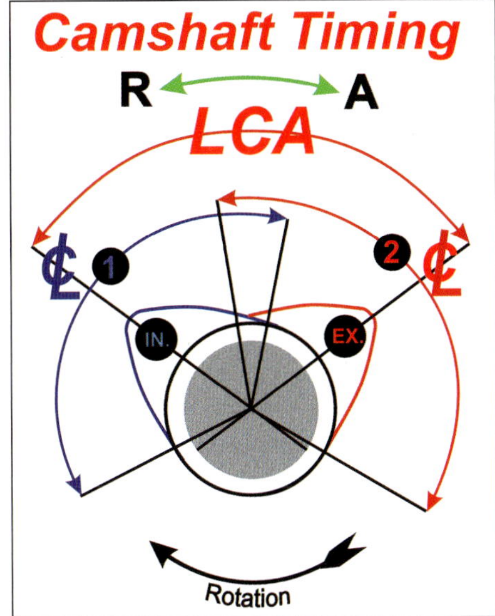

Shown here are the various cam parameters we will be discussing in this chapter. The most important to appreciate are the intake (1) and exhaust center lines (2), and the LCA and its relation to cam advance (A) and retard (R).

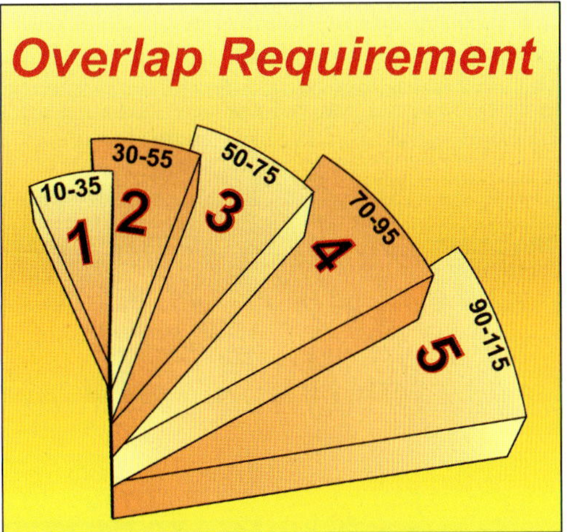

Fig 7-2. In this overlap chart, Section 1 is for working trucks, where low-end torque and mileage is an important requirement. Cams of this duration on the right LCA will provide strong performance right off idle, providing excellent heavy-load towing capability, good mileage, and a smooth idle. Such cams should always be used with high-lift rockers, as the short duration limits the lift that can be applied to the cam lobe.

Section 2 is for daily driven cars and trucks where a strong emphasis is still on idle quality and low-speed performance for good street manners and strong off-the-line response with a stock stall converter.

Section 3 is for hot street machines, where performance is now the primary requirement, but reasonable street manners must still factor into the equation. In this section, it's important to err on the short side rather than over-camming the motor, especially if used in a heavy vehicle.

Section 4 is for oval track, all-out road race, and heavier drag-race cars. Section 5 is for "no compromise" drag-race engines. In these last two sections, using the appropriate compression ratio is an important aspect toward success.

then even a modest street motor should show at least 180 psi if cam and compression are roughly right. However, in my opinion, that is a worst case. For a regular-use performance street motor I'd expect to see 200 psi. For a street/strip motor, this should be 220, and 240 for a race-only engine. Assuming no pressure is lost to leakage, anything below these levels indicates too low a CR. For these reasons, it's important to follow the CR requirement given for each cam in the selection charts. If for some reason the required CR for a cam can't be attained, choose a shorter duration cam; you'll like the result better.

Overlap

I want to make this next point absolutely clear. It is not the selection of a long cam duration that kills drivability (as opposed to low-speed torque) and vacuum, but overlap. Overlap is the time around TDC when both the intake and exhaust valves are open, and it's a function of both duration and lobe centerline angle. Overlap is crucial for obtaining high output from a race engine. Though good for power, it is, if overdone, destructive to low-end output, drivability, and mileage. This makes the decision as to how much should be used for a given application a critical decision. It's tempting to add a few more degrees in the hopes it won't cost much at the bottom end, but in reality, it's like playing Russian roulette with numbers instead of bullets. For example, selecting a cam that has 40 degrees of overlap instead of 30 degrees may only appear to be adding an extra 25 percent, but this isn't the case. The critical factor is the overlap area brought about by the combined degrees and lift. Adding more overlap also adds lift during the overlap period. What appears to be only an extra 25 percent of overlap actually ends up being about 80-percent more. The bottom line is, don't be tempted to select a cam having too much overlap.

The effect overlap through-flow area and cylinder displacement size have on the power curve are closely linked. For a given displacement, the greater the through-flow area is (overlap degrees times lift), the more RPM the engine needs before it starts to perform. The bigger the engine is for a given cylinder head, valve size, or overlap through-flow area, the lower the RPM at which it starts to function. This means the selection of the overlap must be based largely on the lowest RPM that the engine is required to run from.

Fig 7-2 gives an indication of what's required for a small-block Chevy equipped with an exhaust system with near-zero backpressure and an unrestricted intake.

To put numbers to these, we're talking about an exhaust with less than 0.5-inch of mercury (0.25 psi) and less than 1.5 inches (Hg) of intake manifold vacuum at peak power. If exhaust backpressure increases, the amount of overlap must be decreased. If the intake is restricted, the situation becomes complex and is

This is a simple explanation as to how vacuum drops when a single plane intake is used and in conjunction with the cam getting bigger. The induction cycles in a V-8 are 90-degrees apart. That means when one piston is at TDC in the overlap period (left-hand cylinder in this drawing) another is half way down the bore on its induction stroke. The small blue circle represents the area available at the nearly closed carb butterflies. The larger red dots represent the through flow area from the exhaust system through the exhaust valve, into the cylinder, and on out the intake valve

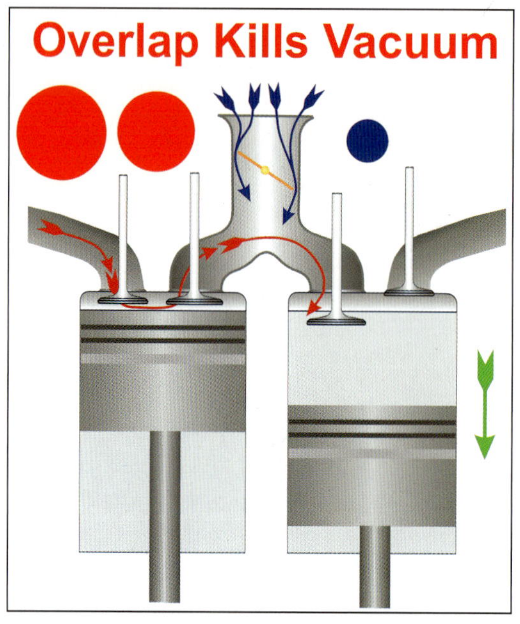

into the intake manifold. The smaller red dot is for a moderate-performance street cam, while the larger is for a big street performance cam. You can see how much easier it is for the cylinder on its induction stroke to draw on exhaust rather than through the carb restriction.

influenced to an even greater extent by the exhaust used. All the cams in the cam selection chart are based on the premise of enough overlap for the job and no more.

LCA Selection

Within both the industry and the hot rodding community at large, the most difficult aspect of cam speccing seems to be determining what LCA will be the most effective. My intention here is to present a basic understanding of what's required so you can get an idea of how the optimal LCA is influenced by the spec of the rest of the engine. In doing so you will also begin to understand why the cam that may have worked so well in your friend's 327 won't be remotely close for your 383.

One basic rule that can be applied across the board is: The more an engine becomes under-valved, the tighter the LCA needs to be. Normally, the LCA required is based on the cylinder head's flow capability and the size of cylinder to be serviced. Unfortunately, only a minority group of builders are in a position to

get cylinder-head flow figures. This means adopting the next best method that produces credible results. As it happens, adopting intake valve size instead of flow gives almost equally good results.

What we look at here is the number of cubic inches each square inch of valve opening area (lift x valve circumference) has to feed. The more cubes per square inch, the tighter the LCA must be. This is why the optimum LCA for a hot 302 is too wide for a 383. Using a cam that worked in the 302 will prevent the increased torque potential of the 383 from being fully realized.

Before finishing with this subject, we need to deal with the effects this angle supposedly has on idle. Many off-the-shelf cams have wider LCAs to preserve idle quality. Many cam companies grind their cams on LCAs a little wider than optimum for maximum output. Intentional or not, this does save the hot-rodders from themselves. How? Because almost all of them work on the Stroker McGurk theory: If some is good and more is better, then too much must be just right. If you're building

a street motor and idle quality is important, you may be pressured from sources other than this book into considering a cam with a wider LCA. However, you will only need to go this route for a quality idle if you've been over-enthusiastic in selecting the overlap to be used in the first place. In other words, a wider LCA than the engine actually needs is a second-grade Band-Aid fix for your enthusiasm in terms of duration selection!

Duration

We are now getting to yet another important factor that relates engine size to cam selection. Read this twice if you have to, but be sure you understand what I say. Adding valve opening duration has always been an accepted technique for making more power upstairs, but it has its pros and cons. Adding duration is easy on the valvetrain and makes for reliability in that department. But the price of added top end is almost a direct tradeoff, because it reduces low-end output. Just how much it reduces low-end output depends on the ratio of valve opening area to cylinder volume. Small-displacement big-valve engines rapidly lose low-end output compared to under-valved larger units. The bottom line is that if you're using a smaller version of an engine, you need to be more conservative with duration. If the engine has been stretched by boring, stroking, or both, a little extra duration proves an asset.

By extending cam timing, we cause the torque curve peak to move up the RPM range. More cam timing doesn't necessarily mean more torque, just more RPM at which it occurs, and this translates into more horsepower. Fig 7-3 shows the effect of more duration while the CR remained the same throughout the test. The cams No. 1 through No. 3 are nominally 270, 285, and 300 degrees of seat duration. It doesn't take much studying of these curves to see how much of a direct tradeoff added duration can be. However, because the CR wasn't increased for the longer cams, these

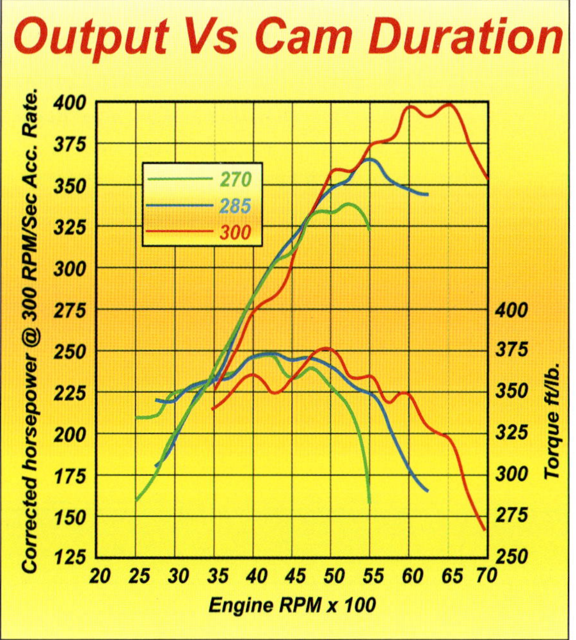

Fig 7-3. It's clear to see that longer duration cams don't necessarily produce any more torque (all else being equal) but produce roughly the same peak torque but at higher RPM. This is where the extra horsepower comes from. To allow a wider useable RPM range, the cams used for this test were on a wider-than-optimum LCA, and the CR wasn't raised as each subsequently longer cam was tested. Had the optimum LCA been used, the torque output of a motor, such as this, would have increased 30 ft-lbs.

What you see here are the components that determine whether or not we have a functionally good valvetrain.

tests show a "worst case" situation. Adding the appropriate CR would redeem half of what was lost at the low end while adding to the top end output.

Before deciding to use a long duration, carefully consider the RPM range you want the engine to have. Also be aware that longer duration cams become fussier as far as the rest of the engine spec is concerned. This means it's far better to err on the short side, especially on a street motor, as you'll like the results far better.

Dual-Pattern Cams

Many cams have different intake-to-exhaust duration. For high-compression high-RPM applications, this is desirable because of the reduced time period within which the exhaust must be expelled. Usually about two-thirds of the extra opening period, usually from four to 10 degrees, is on the opening side of the exhaust, thus giving the cylinder more time to "blow down." On a purpose-built performance street motor, this extra blow down can cut mileage and low-end torque by as much as 5 percent— sometimes more.

From mid-range RPM on, extra exhaust duration starts to help power. Some off-the-shelf street cams are spec'd this way because the cam designer anticipated their use in engines with standard or inadequate exhaust ports and systems. If a true street engine is equipped with a modified head and a free flow exhaust system, a single-pattern cam will often be the best choice. A dual-pattern really starts to pay off when dealing with engines with higher compressions, higher RPM capabilities, and/or nitrous. We won't be dealing with blown or turbo motors in this book, but we will be going into nitrous motors in some detail. A nitrous motor can have significantly different cam requirements.

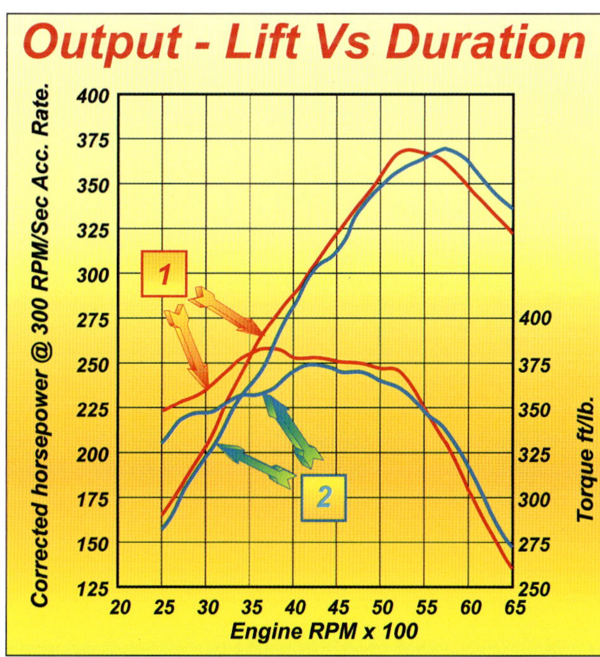

Fig 7-4. For this test, Cam No. 1 was a shorter duration but its faster opening profile in conjunction with high lift rockers produce close to the same area as Cam No. 2. This cam had more duration, but with 1.5 rockers produced less lift. What we learn from this test is that in a motor that's under-valved and RPM-restricted by stock bottom-end components, it's a good idea to go for all the valve lift possible, but be conservative on duration.

Lift Versus Output

The number-one question concerning lift is how much is needed before any further increase becomes pointless. This might seem tricky, but fortunately there is a simple answer for a small-block Chevy. Because flow goes on increasing to high lift, and because the intake valve is too small for the cylinder concerned, we can say with a great deal of certainty, "The more lift the better."

It's revealing to compare the output of two valvetrains (Fig 7-4) generating similar valve opening lift area in terms of inch/degrees.

Although each produces about the same horsepower, the short, high-lift design makes more torque, and as a consequence produces its power with fewer RPM. This means less RPM-induced stress on bottom-end parts. Ultimately, it's RPM that breaks bottom-end parts. A prime goal is to make horsepower without excessive RPM. That means building torque. By making a better choice of valvetrain, we delay the point at which heavier-duty bottom-end parts are needed.

From the curves in Fig 7-4 we can clearly see that cam No. 1 produces the most usable output—especially for street use. The implication is that we should specify a valvetrain that delivers the most lift possible within the dynamic constraints imposed. Unfortunately, building a high-lift/short-duration valvetrain is more difficult than a longer-duration/lower-lift one. Achieving our objective means opening and closing the valve and lifting it higher in a shorter time. We can't just add spring to control the situation because this means inexpensive parts wear too quickly and costly ones are out of our financial reach. To achieve our goal of power on a tight budget, we have to make smarter choices from the less expensive hardware available. That boils down to choosing the right parts and assembling them correctly. Details such as valve, retainer, pushrod, and lifter

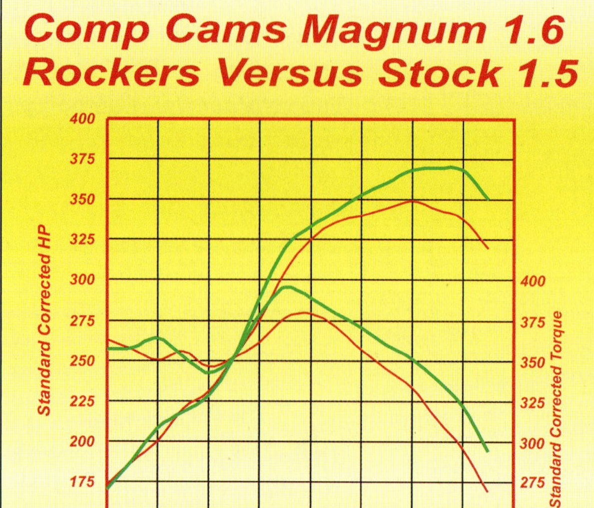

Fig 7-5. The effect that swapping out stock stamped steel rockers for COMP's magnum rockers is shown. The cam used in this test was a stock 929 hydraulic cam, which is on a 112 LCA.

weight are important, but the principle factors are the valvesprings' capability and cam profile dynamics.

For a street or even a semi-race application, the lift numbers for best output are far higher than can be achieved with available hardware, unless we are dealing (against my advice) with a small-displacement engine. Only for opening durations in excess of about 300 degrees does it become vaguely possible to approach maximum desirable lift. This alone should indicate that maximizing lift, especially for a street motor, should be considered a priority.

Assuming you have or are going to have a decent head on your engine, going for a little more duration and as much extra lift as possible is the way to go. For pushrod engines this means a serious look not only at the cam profile, but also the rocker ratio, which is a little low at 1.5:1 on a stock small-block. The amount of lift that can be ground on a cam is limited. The shorter the cam, the less the lift. The slower the acceleration (as in production style cams), the less the lift. Attempting to establish which of two similar duration profiles has the fastest opening is difficult, but in most cases it will be the cam with the most lift, so lift once again becomes a predominant factor.

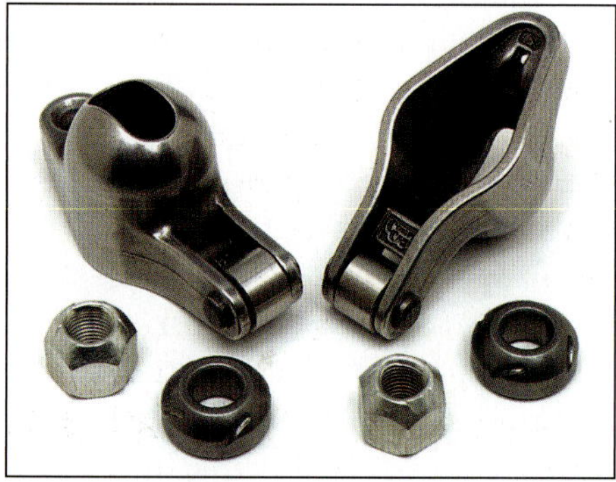

COMP's Magnum rockers represent an excellent entry-level high lift rocker.

Crane's stamped steel 1.6:1 rockers are about as cheap as you can get for higher valve lift.

Since we can't get enough lift into our small-block's valvetrain, we must adopt the highest-ratio rocker our budget allows. Higher-ratio rockers not only produce more lift, but also increase opening

Seen here is the COMP aluminum rocker that I use on most of my builds. Also note the beehive springs with their small-diameter 12-gram retainers.

rates, thus presenting the cylinder with breathing area faster and sooner. This, for the most part, means adopting 1.6:1 rockers. Unfortunately, indiscriminate use of high-lift rockers in conjunction with a high-lift cam and heavier springs can bring about wear problems unless you know what you're doing.

This chapter emphasizes lift as a performance parameter. Also recommended are cams that are on tighter LCAs than usual, although shorter in duration. These factors combined, especially on bigger-inch engines, means that the piston-to-valve clearance is used up quicker. Do yourself a favor: always do a trial check of piston to valve clearance. You do not have to wait until final assembly to do this. Just mount a piston, without the rings, on a rod and install it along with the cam and a head but with just the valves of one cylinder installed. Put some clay on the pistons and turn the assembly over. The minimum clearance you are looking for is 0.080. If it is less than this, have your machine shop cut the piston's valve pockets appropriately. Be aware, though, that most pistons have plenty of clearance for big street cams in 383- to 400-inch engines with the cams I recommend.

Lift Versus Wear

Controlling a higher-lift valvetrain almost inevitably means stronger springs. This, coupled with the greater motion of the valve, means more side loading on the valvestem and guide. This can cause more wear. If a stock cast-iron guide is used, then anything over a 0.5-inch lift and about 270 pounds of over-the-nose spring force can wear production cast-iron guides at an unacceptable rate for the street. For instance, the guides may be well worn to the tune of costing 20 to 30 hp after as few as 30,000 miles (although this situation gets better as oils get better). The fix is bronze guides, optimizing valvetrain geometry and/or roller-tipped rockers, and—most important—keeping the required spring force to a minimum for the job. Installing bronze guides and acquiring roller-tipped rockers is straightforward and needs no further explanation, so let's deal with the budget, valvetrain geometry, and springs.

At this point let's assume that you don't have the cash for new rockers and are forced to use the existing stock ones. This is acceptable if the ball and rocker are in good condition and the tip isn't worn. If the head and block have been decked, the stock pushrods are now too long. This will push the rockers up in relation to the valve. This in turn causes the rocker tip to rotate farther down at the full-lift position, and any wear ridge that was present will now run onto the valve tip and generate noise and extra side thrust. If tip wear is less than 0.005 inch and the budget is tight, you can salvage rockers and replace them at a later date. To make them usable, dress out the wear with emery paper wrapped around a flat file. Don't take off any more metal than necessary so as to retain whatever case hardening may remain.

Also, unless they are the "long slot" variety, you may find that the greater range of rocker motion causes the rocker to run out of slot length and bind against the rocker stud. Check for this and grind the slot longer, as required.

To maximize the tip life of these and, indeed, any non-roller rockers, hardened valve-tip lash caps should be used. These serve two purposes: first, they spread the rocker tip wear over considerably more area, and second, they help compensate for too long a stock pushrod. In setting up the valvetrain geometry, especially with non-roller followers, we're attempting to minimize the rocker's contact patch sweep across the valve tip. Unfortunately, in addition to the sweeping action across the end of the valve there's also a part rolling action due to the radius on the rocker tip. Combining these two aspects and attempting to determine what gives the least side load is far more complex than you might expect. Simple rules such as the half-up/half-down rule that work with roller followers do not deliver in the case of a non-roller follower. This being the case, here are my recommendations for the budget motor with machined heads, block, or both, utilizing non-roller rockers of either 1.5 or 1.6:1 ratio.

First, always use a lash cap. These are usually 0.050 or 0.080 inch, and if the budget allows use of pushrods 0.050- to 0.060-inch shorter for cams up to 0.500- or 0.525-inch lift. Pushrods as much as 0.10-inch shorter may be needed for cams lifting to 0.60-inch, but valvetrains in this category are moving well out of the budget class. Consider it mandatory that anything over 0.550-inch lift should really be using a roller rocker.

If roller tip rockers are to be used, the situation gets easier to deal with. If the pushrod lengths are set as per the recommendations just given, the sweep of the rocker across the valve tip will be acceptable for most combinations of heads and rockers, but it still needs to be checked. This should be done with an adjustable pushrod available from any performance parts house or speed shop.

As far as rockers for a basic budget motor are concerned, I use plain-stamped 1.5s and 1.6s since these are about as cheap as can be had. Sources to check for them are Crane, Summit, and Jegs. If the budget stretches for about two- to two-and-a-half times that of the plain stamped rocker, then I use the PRW or COMP Cams' ball-pivot, roller-tipped stainless Magnum rockers. These are stiff and usually produce more lift than the theoretical ratio they're supposed to. The bottom line is they usually make at least 5-hp more than even the stamped 1.6s. Also, if they're not used with more than the recommended 350 pounds over-the-nose spring force, these rockers are about indestructible.

At about 40-percent more money, my next choice for a fully rollerized rocker is the "Energizer" rockers from Crane. There are other brands, but when the situation prevails, I use super-high-ratio rockers from COMP Cams and Crower. With big-inch engines (383 on up), the only way to go is 1.8 to as high as 2:1 ratio, but that's not within this budget.

Making a Cam Selection

Most publications concerning high performance give you a very basic run-down on cams, to the point where you are sufficiently informed to fall just short of making an effective decision on the cam spec your engine should use. It has been a great frustration over the years to see the public misinformed on this subject time after time by books and magazines alike.

It pays to consider here that making the wrong cam spec decision costs just as much money as the right one. The biggest issue is usually which LCA is used. Getting this wrong by just 2 degrees can typically cost 20 ft-lbs and 20 hp!

Calling a cam company might seem like an easy way to get the right cam, but the bottom line is: calling half-a-dozen cam companies will get you half-a-dozen quite different answers. To avoid this situation I am going to do what no other performance book has ever done—give you what must be as near absolutely fool-proof cam recommendations as is humanly possible.

You are very likely to get some opposition here from so-called experts who will claim that the cam specs don't fall in line with their experience. If it lets you rest any easier with the choice of going with my recommendations here, just ask the person in question if they have tested in excess of 10,000 combinations in small-block Chevys. If they have not then maybe you should be putting your confidence in the cams listed at the end of this chapter.

I teamed up with Lunati to do these. You will find that, unlike most catalogs where you will ponder for half an hour or more over a number of selection possibilities (and still get it wrong eight times out of ten), making a selection can be done by the average 12-year-old with almost zero automotive knowledge. As for the functionality of the selection, it will be as good as if done by, say, Pro Stock engine builder/racer Warren "The Professor" Johnson. So when you are ready to make a physical selection of your cam, just turn to the last few pages of this chapter.

Springs

The spring looks as if it's the simplest part of our valvetrain to get right. In practice, it's the single-most critical and complex component in terms of striking compromises in the whole valvetrain. As far as the performance of the valvetrain is concerned, it's solely governed by the spring's capability. Selecting inappropriate springs will limit what can be done. Too much spring force, especially over the nose, wears out flat-tappet cams fast; too little, leads to loss of control.

The first rule is to minimize the amount of spring force needed to achieve desired RPM. The second is to select the best spring for the job. In case you thought this is easy to figure out, let me set you straight. It's not just a question of selecting a spring that will give the "seat" and "open" forces you think are needed. There isn't room here to go into the science of spring design to any depth, so I'll attempt to keep the theory part short. Ideally, we need a spring that gives the seat and nose forces and has the highest natural resonant frequency of those available to us. What this really boils down to is the selection of a spring with the highest delivered force-to-weight ratio. In so choosing, we're moving the spring's propensity for running into spring surge as far up the RPM range as possible and hopefully out of the range we intend to use.

To give you an example of how important spring mass is, recounting some tests I ran with Brian Crower on Crower's spin test machine will demonstrate the point. The tests involved three springs weighing in at 82, 126, and 142 grams. The poundage seat and nose were 118/342 for the 82-gram spring, 162/405 for the second, and 142/352 for the third. These springs ran the RPM numbers to 8,500, 8,600, and 8,100. As can be seen from the numbers, the RPM achieved, pound for pound, was far better with the 82-gram spring. Why? Because far less of the delivered force is required to control the spring's own mass. This leaves more to control the rest of the valvetrain.

With results like this, it doesn't take much to figure which spring will produce the least frictional loss and the longest cam life. If you have access to a spring tester, here's how you select the spring. Get all the possible candidates that will produce the desired seat and full-lift force. The spring package to use will be the lightest one of the group. This system works fine until we get to advanced levels of valvetrain. At that point it's possible to select a spring that becomes over-stressed in use. This is unlikely to happen in our budget scenario.

The spring's job is to constrain the valvetrain to follow the dictates of the cam profile. There are two forms of loss of valvetrain control. These are shown in Fig 7-6. Lofting the valve at high RPM helps generate more power. Winston Cup

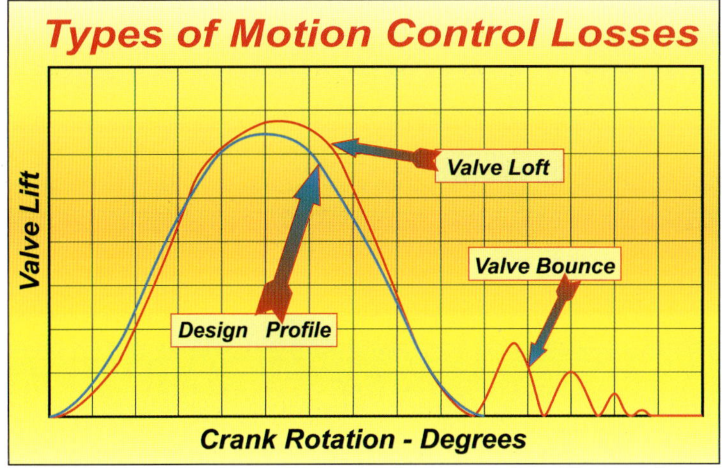

Fig 7-6. Here's what happens as the limiting valvetrain speed is reached. The false motion, known as lofting, valve toss, or valve-train float, is mainly caused by the pushrod flexing, and it can occur as much as 800 rpm before total loss of control takes place at the valveseat. Lofting aids power as it increases the opening area under the lift curve. Seat bounce on the other hand, causes a large drop in output.

Here is what can happen to flat tappet lifters that are not installed using the correct procedures. These lifters have less than 500 miles on them and have not only worn more than 1/8 inch off the lifter face but also a like amount from the cam lobe.

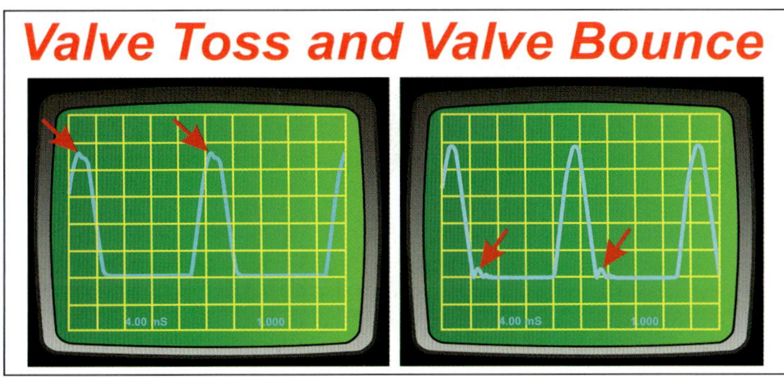

The output from a Spintron test shows the two principle types of motion control loss. Valve toss is on the left and seat bounce on the right.

This is the cam that was paired to the lifters above. Note the lobe on the left is nearly non-existent.

motors rely on valve lofting, which picks up about 0.030- to 0.040-inch lift, for the last 40 hp of their output. Lofting is a secret weapon reserved for solid flat-tappet cams. For hydraulics it leads to what is commonly known as lifter pump up, and for rollers it's a good way to ensure a short roller life.

Around the turn of the millennia a new player appeared on the hot rod scene. This was the beehive spring. Although it had been around a long time, the beehive spring did not find its way to any great extent in the auto industry until GM started using it in large numbers in the early 1980s. In 2002, COMP took a serious look at this spring design for use in its Chevy LS1/LS6 cam line. The resultant spring proved to be a winner. Because the beehive form produced a spring with a far less clear-cut resonant frequency, more of the spring's delivered force is used to control the valvetrain instead of its own mass. The result is less spring force for more RPM.

Although a little more money, the beehive spring is so much better that it really does bear serious consideration. For instance, a relatively short, nitrided, aggressive hydraulic-flat-tappet with high-lift rockers will deliver more torque and power than a more-expensive roller setup with conventional springs (and we are talking cam durations less than about 274 degrees seat-to-seat here).

The subject of lifter pump up was mentioned a few paragraphs back and now seems an appropriate time to talk about it. When lifter pump-up occurs, it's because the lifter is doing exactly what it was designed to do, namely, take up excessive valvetrain clearance. If the valve starts lofting, it's mostly due to pushrod flex. This in turn may cause it to unseat from the tappet, and this separation looks exactly the same as unwanted lash. The lifter responds by increasing in length. Unfortunately, the lifter can get longer a lot quicker than it can shorten. Any added length results in the valves being held off the seats. You do not need an anti-pump-up lifter to fix this. The problem is caused either by a driver who can't read a tach and shift at the correct RPM, by spring surge (resonance), or simply insufficient overall spring force. The fix for each of these is self-evident.

So much for the lifter pump-up diversion. Let's get back to the subject of valve motion problems. Unlike lofting, valveseat bounce is sure-fire death to horsepower. Certain factors affect how we strive to control each type of normally undesirable valvetrain motions. Too

Beehive Spring Comparison

How Much Mass Reduction?

Shown below are the masses involved when a comparison is made of Comp's 987 spring and associated hardware versus their 26918 beehive spring. When the valve opens, not all of the spring moves so the effective mass is not the same as the spring's total weight. Here in the chart we are taking the generally accepted figure of 1/3 of a spring's total mass as being its effective mass.

Component	Mass (grams)	Effective Mass (grams)	Component	Mass (grams)	Effective Mass (grams)
987 Spring	111	37	26918 Spring	74	15
Retainer	28	28	Retainer	11	11
Locks	6	6	Locks	4	4
Valve	123	123	Valve	123	123
Total	268	194	Total	212	153

The difference in size, and consequently weight between a beehive spring assembly and a conventional spring assembly, can clearly be seen here.

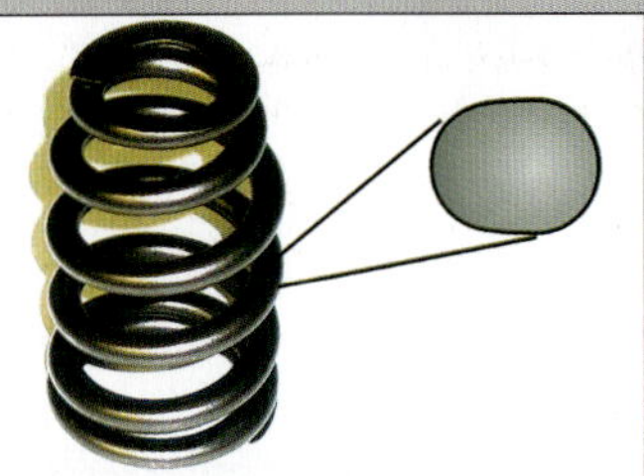

The wire used for a high-tech beehive spring is ovate rather than round because its shape distributes the wire stresses more evenly.

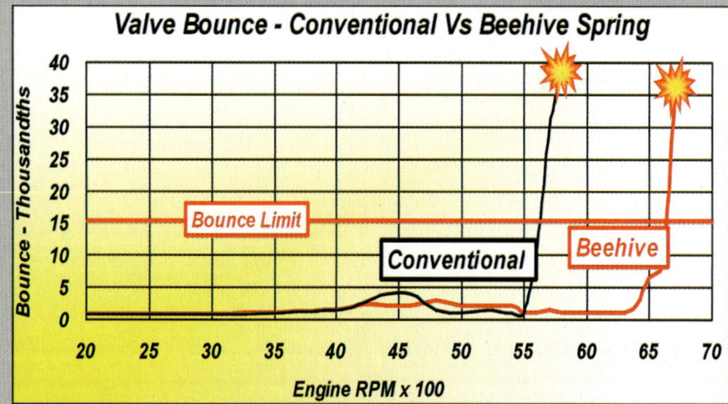

I ran this test at COMP. Note how much better "behaved" the beehive was in terms of seat bounce as the RPM progressed through the operating range.

The total spring assembly's mass is 194 grams for the conventional spring and 153 for the beehive spring. That's about the same reduction as you would see if a conventional spring and an ultra-expensive titanium valve were used. In all, the valve and associated hardware mass is reduced by 21 percent. But, as good as this is, the gains from a beehive don't stop here.

How Much More RPM?

From time to time I get to spend a few days on a valvetrain test rig known as a Spintron. A two-day session on one of these afforded the opportunity to test a bunch of springs, retainers, pushrods, etc., as they all relate to beehive springs versus regular springs.

The tests I ran with the engineers at COMP resulted in the graph above. With 5-percent less load on the seat and 17-percent less over the nose, the beehive setup ran almost 1,000-rpm more (up from 5,750 to 6,700 rpm) before total valvetrain crash set in.

How Much More Seat Control?

From the graph of valvetrain motion shown nearby you can see that valve bounce off the seat can be very destructive to the power output. Any seat bounce that allows over 0.015 inch of bounce is considered detrimental, although even smaller amounts of bounce can show as a power loss.

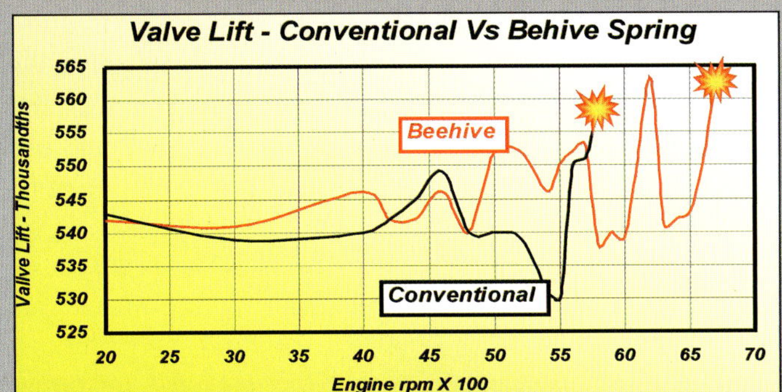

Here is what happens over the cam nose as RPM increases. The static lift on this valvetrain was 0.544. Due to pushrod and rocker flexure, this figure was not reached (left-hand vertical scale on chart) until a small degree of valve lofting started to come about. The greater mass of the conventional spring actually cost lift until the valvetrain started on the final path to loss of control at 5,500 rpm and final crash at 5,700. The beehive allowed for a degree of valve lofting until it lost control at 6,750 rpm.

Because of the reduced mass involved, a beehive spring has far less tendency to allow the valve to "bounce" off the seat. When the use of the spring cures such a problem, the result is a better power curve because spurious dips are less pronounced or even eliminated. In the graph, at left, note that the conventional spring goes through a minor bounce/surge phase at 4,500 rpm but the beehive spring does not.

How Much More Money?

The downside of the beehive springs is that you will pay about twice as much for a high-performance spring as you will for the equivalent conventional dual spring with a damper. However, to get the same RPM you may have to pay just as much for a conventional stronger-spring upgrade as for a beehive. In terms of power output with a hydraulic cam, especially a hydraulic roller, you may not be able to get the performance of a beehive spring even if you paid three times as much for the conventional spring.

Making the Change

The beehive spring is so versatile that just two of the COMP Cams spring P/Ns can replace almost any spring you may currently be using in any small-block cam application from mild street to semi race. Part number 26915 should be used for stock length small-block Chevy valves (approximately 5- to 5.1-inches long) with fitted heights in the 1.75 to 1.8 range and maximum valve lifts of up to about 580 thousandths. For longer valves and or fitted lengths of 1.8 to 1.85 and up to about 615 thousandths lift, use spring P/N 26918. The 11/32 retainers and locks for these are P/N 787-16 and 648-16 respectively. A spring locator is also needed in most instances. P/N 4786-16 gets the job done for every common application seen so far. If the head's spring pocket is smaller than 1.55, the spring locators will need to have their OD ground down. This requires no precision and can be done at home on an off-hand grinder.

much force over the nose not only prevents lofting, which might just be the icing on the cake for a solid flat-tappet cam valvetrain, but also leads to an early demise of the cam nose.

Too much seat load hammers out the valveseats, although it may prevent seat bounce. Too little on the seat will allow seat bounce. To get a working compromise, we need to have the spring rate (stiffness) such that it produces appropriate loads at the valve's closed and open position. At the same time, we should use no more force than necessary. Dealing

with flat-tappet cams, experience indicates that, with regular service and prior to the removal of ZDDP from almost all oil blends (it happened about 2006), at no more than 220 pounds at full lift, a hydraulic flat-tappet cam is good for several hundred-thousand miles. At 240 pounds, about 100,000 miles is all you can count on. If a 75,000-mile life is acceptable, then the nose load can go to about 260 to 270 pounds. At 280 pounds, cam life can drop to 50,000 or less, and predicting gets to be a matter of chance. For an all-out race effort with a regular

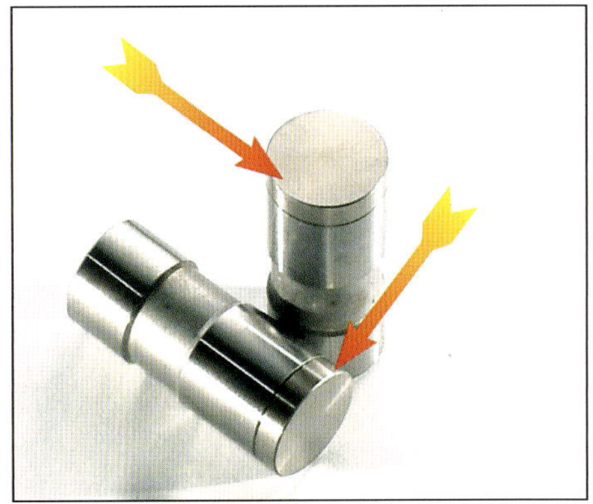

Shown here are the armored lifters I favor. These are almost impossible to wear out!

If we are talking flat-tappet, the cam you select and the lifters used represents 95 percent of the way to reliability. COMP offers a cam nitriding service for flat-tappet cams. I totally recommend going this route.

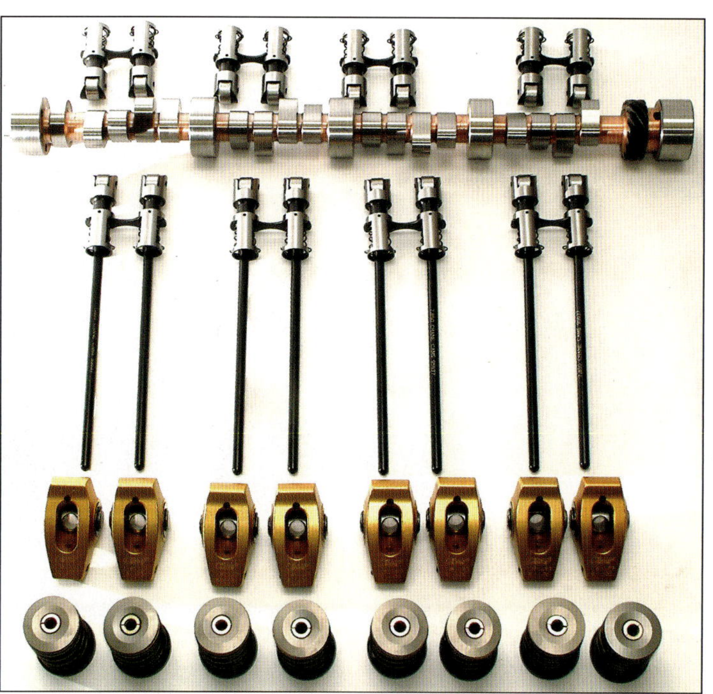

Shown here is a Crane roller valvetrain I successfully used in an 8,800-rpm 350. The 13:1 race motor that used this valvetrain made 618 hp at 7,500 rpm and had a useful power curve to around 8,400 rpm.

cast iron (non hard-faced) flat-tappet cam, consider 350 pounds over the nose as a limit.

So what can be done to improve the chances of a flat-tappet cam's survival? First, you will need to use oils that have ZDDP in them. Most specialty oils do so because they blend to meet the hot rodder's needs, not the EPA's oil standards. A few suggestions here are Amsoil, Quantum Blue, Red Line, Royal Purple, and Torco. (From that list you would think that colored oils are really popular!) You also should check with the cam company for oil additives that address the problem of ZDDP removal from the oils normally sold in auto stores. Another point here is that Lunati (and others) offers a cam nitriding service. This puts a hard layer on the cam lobe and makes for a highly wear-resistant surface. My personal preference here is to always use a nitrided flat-tappet cam and then pair this with hard faced lifters.

For a short cam (less than 270 degrees) motor this makes for a near optimal component combination.

Other than lubrication problems, the most prominent cause of flat-tappet cam lobe failure is spring coil bind. On assembly, make sure the valve can go to full lift and still have at least 0.050-inch more lift available. If the coils clash together during high-speed operation, the springs will lose their force far quicker and may even break. To maintain good spring performance, never float the valvetrain.

Another fix for regular cam wear, although you might not want to hear it, is a less-aggressive profile. In addition to profile selection, cam life also is greatly influenced by the break-in procedure. Use of the recommended cam lube is a must. Also, the lobe nose, which is the point of maximum wear, experiences the highest loads at idle. The first few minutes of running are the most critical aspect of breaking in a new cam, be it street or race. Don't let the motor run below 2,000 rpm for the first 20 minutes. Avoid idling a flat-tappet race cam at low RPM, even if your setup allows it.

If you have opted, as I suggested, for the nitride process on your flat-tappet cam, this will allow as much as 40 to 50 pounds more over the nose without sacrificing life. This means 260 pounds over the nose, even on a more aggressive profile, will last indefinitely in a street motor. For the racer who invests in a quality cam, it's good insurance that the cam will last longer than it needs to, saving the cost of a replacement.

Seat bounce is caused by too high a valve-closing velocity. The best way to deal with it is to set the valve on the seat as gently as possible. In this respect, hydraulic cams can be better than solids, but it's not an open-and-shut deal. That, we'll get to later. Next on the list are tight lash cams having up to 0.020-inch lash clearance. These are often, but not always, better in terms of seat poundage required to prevent seat bounce. Last are the regular lash cams, having clearance from 0.020-inch on up.

Retainers

Retainer weight does make a difference, even on a mild profile. Stock steel retainers weigh in at about 20 to 28 grams. The use of a 13-gram titanium retainer ups the RPM limit of a 6,500-rpm motor about 100 rpm. If the motor ran to 7,800 with steel retainers, the lighter retainer pushes that to about 8,000 rpm. Remember, part of the advantage of a

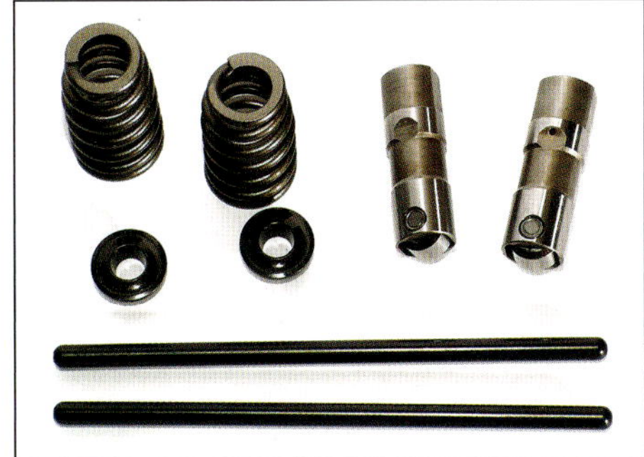

If you are building a hydraulic roller motor, the Components you see here are key elements toward success, especially if RPM over 5,500 is involved.

beehive spring is the 11-gram steel retainer it uses!

Aluminum retainers work as long as they have an extra cross-sectional area to make up for the lower strength. If this is done, they can rival titanium at a much lower price. Be aware that you should avoid over-speeding the valvetrain with aluminum retainers, because they become a problem after such an event more easily than steel.

Lifters

In the order of cost we have the choice of the following lifter options: solid flat tappet, hydraulic flat tappet, solid roller, and hydraulic roller. Unless your engine came equipped with hydraulic rollers, they're out of our price range.

The spring on the right has 5 percent more seat poundage and 17 percent more over the nose than the beehive on the left. The beehive, however, ran over 1,000-rpm more on the Spintron test!

Two things to note here on this Spintron setup. First, see how much smaller the beehive spring is than the conventional spring it replaces. Second, note the use of self-aligning rockers. These ensure rocker alignment and eliminate the need for guide plates.

Short of a non-link bar lifter these Crane roller lifters are the lightest available and as a consequence will turn 80 to 100 more rpm than most other lifters.

These adjustable Isky guide plates will allow the rollers on the rocker tips to be positioned centrally in the side-to-side mode. Another way to achieve alignment is . . .

. . . use self-aligning rockers such as these shown here. By trapping the end of the valvestem between the two hardened washers (arrowed) the rocker is constrained to follow the valve tip.

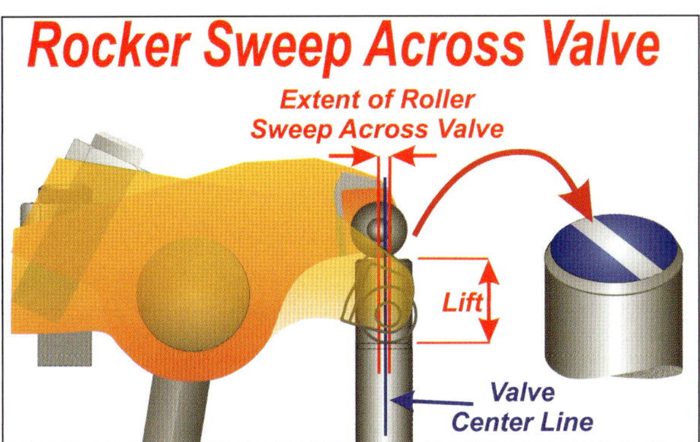

Pushrod length needs to be selected to give this sweep pattern on the tip of the valve. Although I have shown it here in blue, the easiest way is to grease up the end of the valvestem to establish the position of the contact patch.

Solid rollers, however, can just about squeak in if we do our own porting and any other jobs that can, with a little effort, be tackled at home. The most likely and practical choice is a flat-tappet cam. Obviously, the tappet style used reflects the choice of cam. When making a comparison between a solid and a hydraulic cam, do not compare based on seat and 0.050 duration or lift figures, because they won't directly cross over. Because the lash and the checking height are rarely compatible, the duration figures—even at 0.050—can be a little misleading. The easiest way to make a comparison of cams from the same manufacturer is to compare the lift, which for the solid is arrived at by subtracting the required lash. Assuming the same LCA, a solid acts as if it's about 5-degrees shorter than its hydraulic counterpart.

Stock Hydraulic Roller Lifters

If you are rebuilding a post-1987 engine, then it more than likely has hydraulic roller lifters. Although these are fine in a stock application, where RPM much over 5,000 is irrelevant, they can have dire problems with collapse at higher RPM. On the Spintron I have seen as much as 0.100 inch of valve lift just vanish. In practice such a loss not only causes the lift to be inadequate but it also shortens up the duration seen at the valve. Experience on the dyno has shown this can reduce output by anything from (wait for this...it's a shocker) 60 to as much as 120 hp! Unlike flat-tappet hydraulics, it seems the reason for the far-greater hydraulic collapse of roller lifters is that they experience a body-distorting side load during operation. This causes a far greater leakage between the internal components and results in the dramatic collapse seen. The question here is: what can we do to reduce or eliminate hydraulic lifter collapse be it for flat or roller lifters?

The normally recommended technique to set hydraulic lifters is to adjust them until they are into the hydraulic action at about a quarter turn of the rocker adjusting nut. This may be fine for normal use since it ensures indefinite maintenance-free operation, but it is far from the best for high RPM power. Just as with solids, hydraulics can run into valveseat bounce. But in this instance, it can be caused by an increase in closing velocity because the lifter itself has shortened up due to leakage and collapse. If we adjust the lifter almost to the bottom of its travel, there is now little room for it to collapse. Also, the volume of oil needed to refill it is very minimal so its recovery from any collapse is fast. After setting the lifters like this, you need to check the oil flow out of the pushrod because it is possible that one or two lifters out of any batch can restrict oil flow up to the rockers.

If you want to go one step better, then it is entirely practical to strip the lifter and pack it with a small spacer or washers (if you can find any to fit), and limit the plunger travel to about 0.010 to 0.015 inch. With this setup you would lash the lifters about 0.010 from the bottom of the travel.

The first reaction most people have when confronted with this lifter adjustment/modification option is that the lifter will now pump up. The bottom line is that if any lifter pump up is experienced, what it means is that the valvespring is not controlling the valvetrain properly and is pumping up because of component separation. The fix is to re-evaluate the spring used.

Pushrods

A successful pushrod is certainly much more than just a lightweight, stiff component. These are requirements, but the pushrod's ability to remain, as far as possible, unaffected by vibrations is of equal importance. The pushrod will be stiff if the steel is hardened close to its maximum, but it will "ring" far more easily and adverse vibrations will pass back and forth between it and the valvespring.

If it is too soft it will flex and cause loss of controlled motion sooner. Within the structure of what we are trying to do here, a quality 5/16 pushrod will get the job done. Aftermarket chrome-moly pushrods are better and allow more power to be produced if the cam is of the higher intensity type. For all stock, or near-stock applications, a stock pushrod will work but it is a little too long if block and head work has been done.

Timing in the Cam

Here is a real world truth you need to keep well in mind when installing your cam: The more accurately the cam is spec'd, the more important it becomes to time it in accurately. It is easy to see why this must be so.

If we have a cam that is on the wrong LCA, then advancing or retarding the cam could make the intake more effective while making the exhaust less effective, or vise versa. For a number of degrees in either direction, one factor can cancel out the other, so the effect of mistiming the cam is minimal. If all the valve events are on the cam correctly, then mistiming the cam means all the events are off the mark and power will drop much faster. That said, the "correct" cam, timed in a degree or so out, will still make a lot more power than an "incorrect" cam optimally timed.

If the cam is not installed into the engine in the correct position relative to the crankshaft motion, all our work so far will be undone! Surprisingly, there is not a great deal of difference in what is required, whether the engine is near stock or virtually full race. Because of the great similarity of the heads from stock to race in terms of valve-size ratio and intake to exhaust flow ratios, we find that 4 degrees of cam advance pretty much universally hits the mark in a naturally aspirated engine.

OK, I have brought the term "advance" (and by inference "retard") in here. Exactly what defines cam advance? Before going on here just flip back a few pages and re-familiarize yourself with the cam diagram on page 91.

To establish a cam position in the engine, we need a relevant reference figure by which to gauge the amount of advance or retard a cam has when installed in the engine. The reference figure we use is the cam's Lobe Centerline Angle. If the cam is set into the engine to produce full lift at the same number of degrees after TDC on the number-1 inlet valve as the cam's LCA, then the cam is said to be timed straight up. This means it is neither advanced nor retarded. For example, if a cam has a 108-degree LCA and is timed-in so that the intake valve on number-1 cylinder reaches full lift at 108 degrees after TDC,

What you see here is about the Rolls Royce of small-block Chevy adjustable timing gears. Available from COMP Cams it really is a nice piece.

About the cheapest timing set you can get is the link-back set (right rear). These are totally functional, but if the budget stretches to a few more bucks the roller set in front is preferred.

The most basic method to make cam timing changes is to use offset dowel bushings. This involves simply drilling out the cam gear to accept the bushing.

The nearly least-expensive option to effectively correct any cam timing errors is to use a 9-keyway timing set as seen here. The gear is appropriately marked with the number of degrees of advance or retard each keyway delivers.

Here is a Cloys Hex Adjust timing set. I really recommend this setup, for its high-quality chain and gear timing set on the market. Once you have used an adjustable timing set the value of near instant error correction on what is normally a tedious operation will become clearly apparent.

Here are your two options for a cam button to control end float on a roller cam in a pre-1987 block. I mostly use the cheap nylon one on the left and have yet to see one worn out.

Here is a nylon thrust button installed into one of my motors. I try to get the end float to be about zero.

If you want to get a little fancier but still without spending a fortune I recommend this thrust button and reinforced timing cover kit from COMP.

then that is "straight up." If we advance the cam 4 degrees (that is make all the events happen sooner in relation to the crank rotation) the inlet valve would reach full lift at 104 degrees (108 – 4 = 104). If the cam was retarded 4 degrees (all events happen later in relation to the crank rotation) it would produce full lift at 112 degrees after TDC (108 + 4 = 112).

How important is it to get the cam timed in right? It's a job you should do when installing a new cam, but there's a certain amount of leeway within which you can work without losing any significant amount of power.

If you hit the mark within plus or minus 2 degrees, you are unlikely to feel the difference through the seat of your pants, but the drag strip clocks might just show a difference. It depends on whether you were 2-degrees-too-far advanced or 2-degrees-too-far retarded. For the most part, if a cam is too far advanced by a couple of degrees, the change in the power curve over the RPM band used for racing is almost unchanged. The clocks on the strip would show the same whether the cam was right on or 2 advanced. On the other hand, if the cam is 2-degrees retarded, you will see the car take longer to stop the clocks on the big end of the strip. If you want to contain any possible loss of output from a mistimed cam to about 0.3 percent, then figure that the tolerance you need to work within is +2 degrees of advance to –1 degree of retard over whatever is called for.

Before you start final assembly check that the cam sprocket of the timing set you intend to use clears on post-1987 blocks. Some don't and the block will need to be clearanced as indicated by this arrow.

For example, if the cam was supposed to be 104 degrees for 4 degrees of advance it would be OK from 105 degrees (1 degree-retarded over what it is supposed to be) to 102 degrees (2-degrees advanced over what it is supposed to be).

Installing a Roller Cam

If you decide to go with a roller cam setup in a pre-1987 block, then you will need to take care of the cam's end float because it will not have the thrust plate that takes care of that as per post-1987 blocks. Be aware that you can install the earlier cam/timing set in a later block, but not the other way around. For non-roller blocks the cam will need a thrust button that bumps up against the timing cover.

Finally Making a Cam Selection

I went to some pains earlier to make a point on the uncertainty of cam selection the performance community generally faces. Since making a selection is close let me reiterate the points made earlier. Most performance books will tell you enough about cam selection to totally optimize the uncertainty of your choice. I don't intend to do that here. What I have done instead is to team up with Lunati and, using its profile library, spec'd out a range of cams that are as near goof-proof in terms of application optimization as possible. Not only that, but with the way they are laid out there will be no uncertainty as to what you need to use once you have made up your mind exactly what your application is. The profiles I have chosen to use are biased toward reliability rather than outright power. However, because the valve events are more accurate, these cams will almost certainly make more power than a cam with incorrect events and more aggressive profiles. As far as reliability goes, I am reasoning that, since we are all on a relatively tight budget, none of us can afford to have a motor that even vaguely

approaches grenade status. What this means is that these timing numbers, which are the result of thousands of hours of testing, are within about 1.5 percent of optimum 98.5 percent of the time.

Also notice that all the cams are single pattern. That was done for two reasons. First, in this day and age, street cams need to address fuel consumption as well as power. Extending the exhaust timing to get more power above the 4,000-rpm mark costs mileage. Here a single-pattern cam is better. The second reason for all single-pattern cams is to simplify selection. These charts will deliver an optimal cam spec (for a given intake timing) more than 98 selections out of 100 to within less than 2 percent. To put that into perspective, that's at least 20 times more accurate than the cam industry at large. Don't believe me? Try this: call five cam companies for a spec for your motor and see how many different recommendations you get. Can they all be right? With most companies you can do this test from their catalog. Just look and see how many vague suggestions seem to fit your motor requirements. If you are not yet convinced, consider this: in 30 years I have never lost a cam shootout on the dyno against cam companies.

If you are looking for some fine-tuning on the cam specs recommended here in terms of dual-pattern specs, more aggressive lobes, etc., I am planning to put a more comprehensive selection chart on my web site at MotorTecmagazine.net.

Read This Before You Make a Cam Selection

You don't want to make a hasty cam selection and order something that, though it has the right valve event timing, doesn't actually fit the block you have and has to be returned. Here are a few things to tell the tech guy to get a cam that will fit and function as intended.

First, spec the distributor gear. Any cam that is made from a cast-iron or cast-steel blank can run with a stock steel distributor gear. If you opt for a cam that has to be made from steel billet, as per most race solid roller cams, then a bronze gear will be needed for the distributor.

Remember that if you have a post-1987 roller lifter block, then the cam normally required for this has a stepped nose to accommodate a thrust plate. The timing gear bolt-hole pitch circle is also smaller. This setup will not fit the earlier blocks but the earlier cam/timing gear setup can be used in the later block. In such a case, cam end float needs to be controlled by a thrust button as per the early block. Also, there may be some interference between the timing gear and the block. The photo on page 104 shows this and what to grind away for clearance. If you have a roller block, be aware that that the link bar of aftermarket solid or hydraulic rollers may seat down on the block before the roller has actually contacted the cam. A fix here is to grind about 1/16 inch off the offending area or use the later taller lifters.

If you are building a stroker motor there is a real possibility of rod shoulders hitting cam lobe flanks. The usual fix here is to order a reduced base circle cam for added clearance. If you are using the Scat rods that I recommend, there is far more rod-to-cam clearance than with most other rods so this postpones the need for a reduced base circle cam until more stroke or a longer duration cam is involved. I do not like to just go ahead and order a reduced base circle cam until I am sure the setup actually needs it because the cam dynamics are degraded and the contact loads increase when the base circle is reduced. For what it's worth, most 3¾-stroke cranks with Scat rods clear most stock base circle cams.

If you have absorbed all this you should now be in a position to go ahead and use the selection charts that follow. To determine what cam is needed, go across the *Application* row until you hit the column that has your intended application (Quick Street, Fast Street, or what ever). Go down this column until you get to the row that has your engine's displacement (as listed in the column far left). Where the row and the column cross is the cam your engine needs.

One last point here concerns cam duration and engine displacement. As the displacement goes up, the cam acts smaller. For instance a "true street" cam in a 350 will act like a "street and tow" cam in a 400. You can figure that for each 25 to 30 inches over 350-inches displacement, the cam will act as if it is about 6-degrees shorter. This gives you the option of stepping across to the right by one column over what would have been the selection for a 350. Let's do an example here. Say you want the characteristics of a "true street" flat tappet hydraulic cam (perfect idle, great low and mid-range with a respectable top end), then, for a 350, the chart would lead you to a [DV256-08HFL] spec cam. But you have a 383 and with those extra inches you could step across to a "quick street" spec and pretty much expect to get those same characteristics. This is an option that is open to you but don't play the game to excess. It is better to have a cam slightly too small than too big.

Power-Profile Endurance Grinds—Hydraulic Flat Tappet

These Lunati cams are spec'ed to David Vizard valve events, with Lunati moderate acceleration rates (medium intensity). These are intended to deliver long life and are less demanding on valvetrain components. This makes them a very cost effective choice for the price-constrained engine build as less-expensive springs and valvetrain components can safely be used. Profiles in blue are good fuel-mileage cams. Important: Use cam break-in lube part #301 to comply with the warranty. Nitriding is recommended for these profiles (add suffix-NH to part #). Before ordering, read the notes below. To order, check your local speed shop or call Lunati at 662-892-1500 and ask for Order Department.

Application	Street & Tow	True Street	Quick Street	Fast Street	Max Street	Street/Strip	Real Race
Nominal Intake Duration	248	256	266	274	282	290	300
Engine CID	Cam Part #	Cam Part #	Cam Part #	Cam Part #	Cam Part #	Cam Part #	Cam Part #
302	DV248-10HFL	DV256-10HFL	DV266-10HFL	DV274-10HFL	DV282-10HFL	DV290-10HFL	DV300-10HFL
327	DV248-09HL	DV256-09HFL	DV266-09HFL	DV274-09HFL	DV282-09HFL	DV290-09HFL	DV300-09HFL
350	DV248-08HFL	DV256-08HFL	DV266-08HFL	DV274-08HFL	DV282-08HFL	DV290-08HFL	DV300-08HFL
383 (3.75-inch Stroker)	DV248-06HFL	DV256-06HFL	DV266-06HFL	DV274-06HFL	DV282-06HFL	DV290-06HFL	DV300-06HFL
400	DV248-05HFL	DV256-05HFL	DV266-05HFL	DV274-05HFL	DV282-05HFL	DV290-05HFL	DV300-05HFL
434 (4-inch Stroker)	DV248-04HFL	DV256-04HFL	DV266-04HFL	DV274-04HFL	DV282-04HFL	DV290-04HFL	DV300-04HFL
Recommended Min. CR	8.3:1	8.5:1	8.8:1	9:1	9.8:1	10.5:1	11.5:1
Lunati Profile #s	UH7/UH7	UH6/UH6	UH3/UH3	UH25/UH25	UH5/UH5	UH20/UH20	UH21/UH21
Advance	4 Degrees	4 Degrees	4 Degrees	4 Degrees	4 Degrees	4 Degrees	4 Degrees
Duration @ Lash/0.050 – In.	248/199	256/207	266/217	274/226	282/231	290/239	300/244
Duration @ Lash/0.050 – Ex.	248/199	256/207	266/217	274/226	282/231	290/239	300/244
Lobe Lift – In./Ex.	0.267/0.267	0.288/0.288	0.303/0.303	0.312/0.312	0.323/0.323	0.333/0.333	0.335/0.335
Rec. Rocker Ratio – In./Ex.							
302	1.6/1.6	1.6/1.6	1.6/1.6	1.5/1.5	1.5/1.5	1.5/1.5	1.5/1.5
327	1.6/1.6	1.6/1.6	1.6/1.6	1.6/1.6	1.6/1.6	1.6/1.6	1.6/1.6
350	1.6/1.6	1.6/1.6	1.6/1.6	1.6/1.6	1.6/1.6	1.6/1.6	1.6/1.6
383 (3.75-inch Stroker)	1.65/1.65	1.65/1.65	1.65/1.65	1.65/1.65	1.65/1.65	1.65/1.65	1.65/1.65
400	1.65/1.65	1.65/1.65	1.65/1.65	1.65/1.65	1.65/1.65	1.65/1.65	1.65/1.65
434 (4-inch Stroker)	1.65/1.65	1.65/1.65	1.65/1.65	1.65/1.65	1.65/1.65	1.65/1.65	1.65/1.65
V/Lift w/ Rec. RR (1.5/1.6/1.65)							
Valve Lift – In.	0.428/0.441	0.461/0.476	0.484/0.499	0.468/0.499/0.515	0.485/0.517/0.533	0.500/0.541/0.558	0.503/0.536/0.553
Valve Lift – Ex.	0.428/0.441	0.461/0.476	0.484/0.499	0.468/0.499/0.515	0.485/0.517/0.533	0.500/0.541/0.558	0.503/0.536/0.553
Recommended Spring #	73943	73943	73943	73943	73943	73943	73943
Retainer (Steel)	75742LUN	75742LUN	75742LUN	75742LUN	75742LUN	75742LUN	75742LUN
Locks	77003	77003	77003	77003	77003	77003	77003
Recommended Lifter #	71817PR	71817PR	71817PR	71817PR	71817PR	71817PR	71817PR

Notes: **Engines with strokes greater than 3.75 may need a reduced base circle cam for rod clearance (add –R to part number).** Post-1987 factory hydraulic roller blocks need the appropriate cam snout to accommodate the thrust plate. Be sure to state post-1987 roller block when ordering. The timing chain cam gear also needs to suit this later cam. These cams are intended for engines with intake valves between 2.02 and 2.08 inches diameter. If 1.65 aluminum rockers exceed budget, they can be replaced with lower priced 1.6 rockers. A reduction of 5 to 7 hp can result. On engines capable of turning more than 6,000 rpm the valve springs, retainers, and keepers can be upgraded to a beehive spring setup. More info at Motortecmagazine.com.
© Copyright and intellectual property of David Vizard.

Super-Profile Hi-Po Street & Strip Grinds—Hydraulic Flat Tappet

These Lunati cams are spec'ed to David Vizard valve events and are designed with more aggressive profiles (high intensity) than the Endurance grinds. As such, they call for a moderate upgrade in spring spec and associated hardware. This makes them a reliable choice for the street performance enthusiast who wants to maximize output while preserving reliability and containing costs. Profiles in blue are good fuel-mileage cams. Important: Use cam break-in lube part #301 to comply with warranty. Nitriding is strongly recommended for these profiles (add suffix -NH to part #). Before ordering, read the notes below. To order, check your local speed shop or call Lunati at 662-892-1500 and ask for order dept.

Application	Street & Tow	True Street	Quick Street	Fast Street	Max Street	Street Strip	Real Race
Nominal Intake Duration	250	262	268	276	280	288	296
Engine CID	Cam Part #	Cam Part #	Cam Part #	Cam Part #	Cam Part #	Cam Part #	Cam Part #
302	DV250-10HFH	DV262-10HFH	DV268-10HFH	DV276-10HFH	DV280-10HFH	DV288-10HFH	DV296-10HFH
327	DV250-09HFH	DV262-09HFH	DV268-09HFH	DV276-09HFH	DV280-09HFH	DV288-09HFH	DV296-09HFH
350	DV250-08HFH	DV262-08HFH	DV268-08HFH	DV276-08HFH	DV280-08HFH	DV288-08HFH	DV296-08HFH
383 (3.75-inch Stroker)	DV250-06HFH	DV262-06HFH	DV268-06HFH	DV276-06HFH	DV280-06HFH	DV288-06HFH	DV296-06HFH
400	DV250-05HFH	DV262-05HFH	DV268-05HFH	DV276-05HFH	DV280-05HFH	DV288-05HFH	DV296-05HFH
434 (4-inch Stroker)	DV250-04HFH	DV262-04HFH	DV268-04HFH	DV276-04HFH	DV280-04HFH	DV288-04HRH	DV296-04HFH
Recommended Min. CR	8.3:1	8.5:1	8.8:1	9:1	9.8:1	10.5:1	11.5:1
Lunati Profile #s	VH30/VH30	VH32/VH32	VH33/VH33	VH34/VH34	VH48/VH48	VH49/VH49	VH50/VH50
Advance	4 Degrees	4 Degrees	4 Degrees	4 Degrees	4 Degrees	4 Degrees	4 Degrees
Duration @ Lash/0.050 – In.	250/207	262/219	268/224	276/233	280/237	288/245	296/253
Duration @ Lash/0.050 – Ex.	250/207	262/219	268/224	276/233	280/237	288/245	296/253
Lobe Lift – In./Ex.	0.291/0.291	0.312/0.312	0.326/0.326	0.337/0.337	0.344/0.344	0.359/0.359	0.371/0.371
Rec. Rocker Ratio – In./Ex.							
302	1.6/1.6	1.6/1.6	1.6/1.6	1.5/1.5	1.5/1.5	1.5/1.5	1.5/1.5
327	1.6/1.6	1.6/1.6	1.6/1.6	1.6/1.6	1.6/1.6	1.6/1.6	1.6/1.6
350	1.6/1.6	1.6/1.6	1.6/1.6	1.6/1.6	1.6/1.6	1.6/1.6	1.6/1.6
383 (3.75-inch Stroker)	1.65/1.65	1.65/1.65	1.65/1.65	1.65/1.65	1.65/1.65	1.65/1.65	1.65/1.65
400	1.65/1.65	1.65/1.65	1.65/1.65	1.65/1.65	1.65/1.65	1.65/1.65	1.65/1.65
434 (4-inch Stroker)	1.65/1.65	1.65/1.65	1.65/1.65	1.65/1.65	1.65/1.65	1.65/1.65	1.65/1.65
V/Lift W/ Rec. RR (1.5/1.6/1.65)							
Valve Lift – In.	0.466/0.480	0.499/0.515	0.522/0.538	0.505/0.539/0.556	0.516/0.550/0.568	0.538/0.574/0.592	0.556/0.594/0.612
Valve Lift – Ex.	0.466/0.480	0.499/0.515	0.522/0.538	0.505/0.539/0.556	0.516/0.550/0.568	0.538/0.574/0.592	0.556/0.594/0.612
Recommended Spring #	73943	73943	73943	73943	73100	73100	73100
Retainer (Steel)	75742LUN	75742LUN	75742LUN	75742LUN	75740LUN	75740LUN	75740LUN
Locks	77003	77003	77003	77003	77111LUN	77111LUN	77111LUN
Recommended Lifter #	71817PR	71817PR	71817PR	71817PR	71817PR	71817PR	71817PR

Notes: **Engines with strokes greater than 3.75 may need a reduced base circle cam for rod clearance (add –R to part number).** Post-1987 factory hydraulic roller blocks need the appropriate cam snout to accommodate the thrust plate. Be sure to state post-1987 roller block when ordering. The timing chain cam gear also needs to suit this later cam. These cams are intended for engines with intake valves between 2.02 and 2.08 inches diameter. If 1.65 aluminum rockers exceed budget, they can be replaced with lower priced 1.6 rockers. A reduction of 5 to 7 hp can result. On engines capable of turning more than 6,000 rpm the valve springs, retainers, and keepers can be upgraded to a beehive spring setup. More info at Motortecmagazine.com.
© Copyright and intellectual property of David Vizard.

Power-Profile Endurance Grinds—Solid Flat Tappet

These Lunati cams are spec'ed to David Vizard valve events, with Lunati moderate acceleration rates (medium intensity). These are intended to deliver long life and are less demanding on valvetrain components. This makes them a very cost effective choice for the price-constrained engine build as less-expensive springs and valvetrain components can safely be used. Important: Use cam break-in lube part #301 to comply with warranty. Nitriding is recommended for these profiles (add suffix-NH to part #). Before ordering, read the notes below. To order, check your local speed shop or call Lunati at 662-892-1500 and ask for Order Department.

Application	Fast Street	Max Street	Street Strip	Race	Super Race
Intake Duration @ Lash	270	278	290	296	306
Engine CID	Cam Part #	Cam Part #	Cam Part #	Cam Part #	Cam Part #
302	DV270-10FSL	DV278-10FSL	DV290-10FSL	DV296-10FSL	DV306-10FSL
327	DV270-09FSL	DV278-09FSL	DV290-09FSL	DV296-09FSL	DV306-09FSL
350	DV270-08FSL	DV278-08FSL	DV290-08FSL	DV296-08FSL	DV306-08FSL
383 (3.75-inch Stroker)	DV270-06FSL	DV278-06FSL	DV290-06FSL	DV296-06FSL	DV306-06FSL
400	DV270-05FSL	DV278-05FSL	DV290-05FSL	DV296-05FSL	DV306-05FSL
434 (4-inch Stroker)	DV270-04FSL	DV278-04FSL	DV290-04FSL	DV296-04FSL	DV306-04FSL
Recommended Min. CR	8.8:1	9:1	10:1	11:1	12:1
Lunati Profile #s	UF17/UF17	UF14/UF14	UF9/UF9	UF16/UF16	UF2/UF2
Advance	4 Degrees	4 Degrees	4 Degrees	4 Degrees	4 Degrees
Duration @ 0.020/0.050 – In.	262/233	264/232	276/243	282/248	292/255
Duration @ 0.020/0.050 – Ex.	262/233	264/232	276/243	282/248	292/255
Lobe Lift – In./Ex.	0.302/0.302	0.313/0.313	0.345/0.345	0.335/0.335	0.345/0.345
Rec. Rocker Ratio – In./Ex.					
302	1.6/1.6	1.6/1.6	1.5/1.5	1.6/1.6	1.5/1.5
327	1.6/1.6	1.6/1.6	1.6/1.6	1.6/1.6	1.6/1.6
350	1.6/1.6	1.6/1.6	1.6/1.6	1.65/1.65	1.6/1.6
383 (3.75-inch Stroker)	1.65/1.65	1.65/1.65	1.65/1.65	1.65/1.65	1.65/1.65
400	1.65/1.65	1.65/1.65	1.65/1.65	1.65/1.65	1.65/1.65
434 (4-inch Stroker)	1.65/1.65	1.65/1.65	1.65/1.65	1.65/1.65	1.65/1.65
V/Lift w/ Rec. RR (1.5/1.6/1.65)					
Gross Valve Lift – In.	0.485/0.500	0.501/0.516	0.518/0.551/0.569	0.536/0.554	0.518/0.551/0.569
Gross Valve Lift – Ex.	0.485/0.500	0.501/0.516	0.518/0.551/0.569	0.536/0.554	0.518/0.551/0.569
Recommended Spring #	73943	73943	73943	73943	73943
Retainer (Steel)	75742LUN	75742LUN	75742LUN	75742LUN	75742LUN
Locks	77003	77003	77003	77003	77003
Recommended Lifter #	70992	70992	70992	70992	70992

Notes: **Engines with strokes greater than 3.75 may need a reduced base circle cam for rod clearance (add –R to part number).** Post-1987 factory hydraulic roller blocks need the appropriate cam snout to accommodate the thrust plate. Be sure to state post-1987 roller block when ordering. The timing chain cam gear also needs to suit this later cam. These cams are intended for engines with intake valves between 2.02 and 2.08 inches diameter. If 1.65 aluminum rockers exceed budget, they can be replaced with lower priced 1.6 rockers. A reduction of 5 to 7 hp can result. On engines capable of turning more than 6,000 rpm the valve springs, retainers, and keepers can be upgraded to a beehive spring setup. More info at Motortecmagazine.com.
© Copyright and intellectual property of David Vizard.

Super-Profile Hi-Po Street Grinds—Solid Flat Tappet

These Lunati cams are spec'ed to David Vizard valve events and are designed with more aggressive profiles (high intensity) than our Endurance Grinds. As such, they require a moderate upgrade in spring spec and associated hardware. This makes them a reliable choice for the street performance enthusiast who wants to maximize output while preserving reliability and containing costs. Important: Use cam break-in lube part #301 to comply with warranty. Nitriding is strongly recommended for these profiles (add suffix -NH to part #). Before ordering, read the notes below. To order, check your local speed shop or call Lunati at 662-892-1500 and ask for Order Department.

Application	Fast Street	Max Street	Street Strip	Race	Super Race
Intake Duration @ Lash	274	282	290	298	306
Engine CID	Cam Part #	Cam Part #	Cam Part #	Cam Part #	Cam Part #
302	DV274-10FSH	DV282-10FSH	DV290-10FSH	DV298-10FSH	DV306-10FSH
327	DV274-09FSH	DV282-09FSH	DV290-09FSH	DV298-09FSH	DV306-09FSH
350	DV274-08FSH	DV282-08FSH	DV290-08FSH	DV298-08FSH	DV306-08FSH
383 (3.75-inch Stroker)	DV274-06FSH	DV282-06FSH	DV290-06FSH	DV298-06FSH	DV306-06FSH
400	DV274-05FSH	DV282-05FSH	DV290-05FSH	DV298-05FSH	DV306-05FSH
434 (4-inch Stroker)	DV274-04FSH	DV282-04FSH	DV290-04FSH	DV298-04FSH	DV306-04FSH
Recommended Min. CR	8.8:1	9:1	10.2:1	12:1	13:1
Lunati Profile #s	VS10/VS10	VS12/VS12	VS14/VS14	VS16/VS16	VS18/VS18
Advance	4 Degrees	4 Degrees	4 Degrees	4 Degrees	4 Degrees
Duration @ 0.020/0.050 – In.	262/233	270/241	278/249	286/257	294/265
Duration @ 0.020/0.050 – Ex.	262/233	270/241	278/249	286/257	294/265
Lobe Lift – In./Ex.	0.337/0.337	0.347/0.347	0.360/0.360	0.373/0.373	0.387/0.387
Rec. Rocker Ratio – In./Ex.					
302	1.5/1.5	1.5/1.5	1.5/1.5	1.5/1.5	1.5/1.5
327	1.6/1.6	1.6/1.6	1.6/1.6	1.6/1.6	1.6/1.6
350	1.6/1.6	1.6/1.6	1.6/1.6	1.6/1.6	1.6/1.6
383 (3.75-inch Stroker)	1.65/1.65	1.65/1.65	1.65/1.65	1.65/1.65	1.65/1.65
400	1.65/1.65	1.65/1.65	1.65/1.65	1.65/1.65	1.65/1.65
434 (4-inch Stroker)	1.65/1.65	1.65/1.65	1.65/1.65	1.65/1.65	1.65/1.65
V/Lift W/ Rec. RR (1.5/1.6/1.65)					
Gross Valve Lift – In.	0.505/0.539/0.556	0.520/0.555/0.572	0.540/0.576/0.594	0.559/0.597/0.615	0.580/0.619/0.638
Gross Valve Lift – Ex.	0.505/0.539/0.556	0.520/0.555/0.572	0.540/0.576/0.594	0.559/0.597/0.615	0.580/0.619/0.638
Recommended Spring #	73100	73100	73100	73100	73100
Retainer (Steel)	75740LUN	75740LUN	75740LUN	75740LUN	75740LUN
Locks	77111LUN	77111LUN	77111LUN	77111LUN	77111LUN
Recommended Lifter #	70992	70992	70992	70992	70992

Notes: **Engines with strokes greater than 3.75 may need a reduced base circle cam for rod clearance (add –R to part number).** Post-1987 factory hydraulic roller blocks need the appropriate cam snout to accommodate the thrust plate. Be sure to state post-1987 roller block when ordering. The timing chain cam gear also needs to suit this later cam. These cams are intended for engines with intake valves between 2.02 and 2.08 inches diameter. If 1.65 aluminum rockers exceed budget, they can be replaced with lower priced 1.6 rockers. A reduction of 5 to 7 hp can result. On engines capable of turning more than 6,000 rpm the valve springs, retainers, and keepers can be upgraded to a beehive spring setup. More info at Motortecmagazine.com.
© Copyright and intellectual property of David Vizard.

Power-Profile Endurance Grinds—Hydraulic Roller Tappet

These Lunati cams are spec'ed to David Vizard valve events, with Lunati's moderate acceleration rates (medium hydraulic intensity). These are intended to deliver long life and are less demanding on valvetrain components. This makes them a very cost effective choice for the price constrained engine build as less expensive springs and valvetrain components can safely be used. Profiles in blue are good fuel-mileage cams. Before ordering, read the notes below. To order, check your local speed shop or call Lunati at 662-892-1500 and ask for Order Department.

Application	Street & Tow	True Street	Quick Street	Fast Street	Max Street	Street Strip	Real Race
Nominal Intake Duration	258	268	276	280	291	296	304
Engine CID	Cam Part #	Cam Part #	Cam Part #	Cam Part #	Cam Part #	Cam Part #	Cam Part #
302	DV258-10HRH	DV268-10HRH	DV276-10HRH	DV280-10HRH	DV291-10HRH	DV296-10HRH	DV304-10HRH
327	DV258-09HRH	DV268-09HRH	DV276-09HRH	DV280-09HRH	DV291-09HRH	DV296-09HRH	DV304-09HRH
350	DV258-08HRH	DV268-08HRH	DV276-08HRH	DV280-08HRH	DV291-08HRH	DV296-08HRH	DV304-08HRH
383 (3.75-inch Stroker)	DV258-06HRH	DV268-06HRH	DV276-06HRH	DV280-06HRH	DV291-06HRH	DV296-06HRH	DV304-06HRH
400	DV258-05HRH	DV268-05HRH	DV276-05HRH	DV280-05HRH	DV291-05HRH	DV296-05HRH	DV304-05HRH
434 (4-inch Stroker)	DV258-04HRH	DV268-04HRH	DV276-04HRH	DV280-04HRH	DV291-04HRH	DV296-04HRH	DV304-04HRH
Recommended Min. CR	8.3:1	8.5:1	8.8:1	9:1	9.8:1	10.5:1	12:1
Lunati Profile #s	B281A/B281A	B281B/B281B	B389A/B389A	R221C/R221C	R232F/R232F	RF23F/RF23F	R245C/R245C
Advance	4 Degrees	4 Degrees	4 Degrees	4 Degrees	4 Degrees	4 Degrees	4 Degrees
Duration @ Lash/0.050 – In.	258/206	268/215	276/218	280/221	291/232	296/237	304/245
Duration @ Lash/0.050 – Ex.	258/206	268/215	276/218	280/221	291/232	296/237	304/245
Lobe Lift – In./Ex.	0.313	0.326	0.335	0.339	0.333	0.333	0.360
Rec. Rocker Ratio – In./Ex.							
302	1.6/1.6	1.6/1.6	1.6/1.6	1.5/1.5	1.5/1.5	1.5/1.5	1.5/1.5
327	1.6/1.6	1.6/1.6	1.6/1.6	1.6/1.6	1.6/1.6	1.6/1.6	1.6/1.6
350	1.6/1.6	1.6/1.6	1.6/1.6	1.6/1.6	1.6/1.6	1.6/1.6	1.6/1.6
383 (3.75-inch Stroker)	1.65/1.65	1.65/1.65	1.65/1.65	1.65/1.65	1.65/1.65	1.65/1.65	1.65/1.65
400	1.65/1.65	1.65/1.65	1.65/1.65	1.65/1.65	1.65/1.65	1.65/1.65	1.65/1.65
434 (4-inch Stroker)	1.65/1.65	1.65/1.65	1.65/1.65	1.65/1.65	1.65/1.65	1.65/1.65	1.65/1.65
V/Lift w/ Rec. RR (1.5/1.6/1.65)							
Valve Lift – In.	0.501/0.516	0.522/0.538	0.536/0.553	0.509/0.542/0.559	0.500/0.533/0.549	0.500/0.533/0.549	0.540/0.576/0.594
Valve Lift – Ex.	0.501/0.516	0.522/0.538	0.536/0.553	0.509/0.542/0.559	0.500/0.533/0.549	0.500/0.533/0.549	0.540/0.576/0.594
Recommended Spring #	73943	73943	73943	73943	73943	73100	73100
Retainer (Steel)	75742LUN	75742LUN	75742LUN	75742LUN	75742LUN	73943	75740LUN
Locks	77003	77003	77003	77003	77003	75742LUN	77111LUN
(Pre-87 blocks) Rec. Lifter #	72430LUN	72430LUN	72430LUN	72430LUN	72430LUN	77003	72430LUN
(Post-87 blocks) Rec. Lifter #	72910	72910	72910	72910	72910	72910	72910

Notes: **Engines with strokes greater than 3.75 may need a reduced base circle cam for rod clearance (add –R to part number).** Post-1987 factory hydraulic roller blocks need the appropriate cam snout to accommodate the thrust plate. Be sure to state post-1987 roller block when ordering. The timing chain cam gear also needs to suit this later cam. These cams are intended for engines with intake valves between 2.02 and 2.08 inches diameter. If 1.65 aluminum rockers exceed budget, they can be

Super-Profile Hi-Po Street & Strip Grinds–Hydraulic Roller Tappet

These Lunati profiles are spec'ed to David Vizard valve events and are designed with more aggressive profiles (high intensity) than our Endurance Grinds. As such, they can call for higher spec springs and associated hardware. This makes them a reliable choice for the street performance enthusiast who wants to maximize output while preserving reliability and containing costs. Profiles in blue are good fuel-mileage cams. Important: Use cam break-in lube part #301 to comply with warranty. Before ordering, read the notes below. To order, check your local speed shop or call Lunati at 662-892-1500 and ask for Order Department.

Application	Street & Tow	True Street	Quick Street	Fast Street	Max Street	Street Strip	Real Race
Nominal Intake Duration	262	270	278	282	290	294	302
Engine CID	Cam Part #	Cam Part #	Cam Part #	Cam Part #	Cam Part #	Cam Part #	Cam Part #
302	DV262-10HRH	DV270-10HRH	DV278-10HRH	DV282-10HRH	DV290-10HRH	DV294-10HRH	DV302-10HRH
327	DV262-09HRH	DV270-09HRH	DV278-09HRH	DV282-09HRH	DV290-09HRH	DV294-09HRH	DV302-09HRH
350	DV262-08HRH	DV270-08HRH	DV278-08HRH	DV282-08HRH	DV290-08HRH	DV294-08HRH	DV302-08HRH
383 (3.75-inch Stroker)	DV262-06HRH	DV270-06HRH	DV278-06HRH	DV282-06HRH	DV290-06HRH	DV294-06HRH	DV302-06HRH
400	DV262-05HRH	DV270-05HRH	DV278-05HRH	DV282-05HRH	DV290-05HRH	DV294-05HRH	DV302-05HRH
434 (4-inch Stroker)	DV262-04HRH	DV270-04HRH	DV278-04HRH	DV282-04HRH	DV290-04HRH	DV294-04HRH	DV302-04HRH
Recommended Min. CR	8.3:1	8.5:1	8.8:1	9:1	9.8:1	10.5:1	12:1
Lunati Profile #s	VHR-10/VHR-10	VHR-12/VHR-12	VHR-14/VHR-14	VHR-16/VHR-16	VHR-18/VHR-18	VHR-20/VHR-20	VHR-22/VHR-22
Advance	4 Degrees	4 Degrees	4 Degrees	4 Degrees	4 Degrees	4 Degrees	4 Degrees
Duration @ Lash/0.050 – In.	262/211	270/219	278/227	282/231	290/239	294/243	302/251
Duration @ Lash/0.050 – Ex.	262/211	270/219	278/227	282/231	290/239	294/243	302/251
Lobe Lift – In./Ex.	0.338	0.343	0.353	0.357	0.367	0.373	0.377
Rec. Rocker Ratio – In./Ex.							
302	1.6/1.6	1.6/1.6	1.6/1.6	1.5/1.5	1.5/1.5	1.5/1.5	1.5/1.5
327	1.6/1.6	1.6/1.6	1.6/1.6	1.6/1.6	1.6/1.6	1.6/1.6	1.6/1.6
350	1.6/1.6	1.6/1.6	1.6/1.6	1.6/1.6	1.6/1.6	1.6/1.6	1.6/1.6
383 (3.75-inch Stroker)	1.65/1.65	1.65/1.65	1.65/1.65	1.65/1.65	1.65/1.65	1.65/1.65	1.65/1.65
400	1.65/1.65	1.65/1.65	1.65/1.65	1.65/1.65	1.65/1.65	1.65/1.65	1.65/1.65
434 (4-inch Stroker)	1.65/1.65	1.65/1.65	1.65/1.65	1.65/1.65	1.65/1.65	1.65/1.65	1.65/1.65
V/Lift w/ Rec. RR (1.5/1.6/1.65)							
Valve Lift – In.	0.540/0.558	0.549/0.566	0.565/0.582	0.535/0.571/0.589	0.550/0.587/0.605	0.559/0.597/0.615	0.565/0.603/0.622
Valve Lift – Ex.	0.540/0.558	0.549/0.566	0.565/0.582	0.535/0.571/0.589	0.550/0.587/0.605	0.559/0.597/0.615	0.565/0.603/0.622
Recommended Spring #	73943	73100	73100	73100	73100	73100	73100
Retainer (Steel)	75742LUN	75740LUN	75740LUN	75740LUN	75740LUN	75740LUN	75740LUN
Locks	77003	77111LUN	77111LUN	77111LUN	77111LUN	77111LUN	77111LUN
(Pre-87 blocks) Rec. Lifter #	72430LUN	72430LUN	72430LUN	72430LUN	72430LUN	72430LUN	72430LUN
(Post-87 blocks) Rec. Lifter #	72910	72910	72910	72910	72910	72910	72910

Notes: **Engines with strokes greater than 3.75 may need a reduced base circle cam for rod clearance (add –R to part number).** Post-1987 factory hydraulic roller blocks need the appropriate cam snout to accommodate the thrust plate. Be sure to state post-1987 roller block when ordering. The timing chain cam gear also needs to suit this later cam. These cams are intended for engines with intake valves between 2.02 and 2.08 inches diameter. If 1.65 aluminum rockers exceed budget, they can be

Super-Profile Hi-Po Street & Strip Grinds–Solid Roller Tappet

These Lunati profiles are spec'ed to David Vizard valve events and are designed with more aggressive profiles (high intensity) than the Endurance Grinds. As such, these cams can call for higher spec springs and associated hardware. This makes them a reliable choice for the street performance enthusiast who is looking to maximize output while preserving reliability and cost containment. Before ordering, read the notes below. To order these cams, either check with your local speed shop or call Lunati at 662-892-1500 and ask for the Order Department.

Application	Quick Street	Fast Street	Max Street	Street Strip	Real Race
Nominal Intake Duration	275	281	287	293	299
Engine CID	Cam Part #	Cam Part #	Cam Part #	Cam Part #	Cam Part #
302	DV275-10SRH	DV281-10SRH	DV287-10SRH	DV293-10SRH	DV299-10HRH
327	DV275-09SRH	DV281-09SRH	DV287-09SRH	DV293-09SRH	DV299-09HRH
350	DV275-08SRH	DV281-08SRH	DV287-08SRH	DV293-08SRH	DV299-08HRH
383 (3.75-inch Stroker)	DV275-06SRH	DV281-06SRH	DV287-06SRH	DV293-06SRH	DV299-06HRH
400	DV275-05SRH	DV281-05SRH	DV287-05SRH	DV293-05SRH	DV299-05HRH
434 (4-inch Stroker)	DV275-04HRH	DV281-04SRH	DV287-04SRH	DV294-04SRH	DV299-04HRH
Recommended Min. CR	8.8:1	9:1	9.8:1	10.5:1	12:1
Lunati Profile #s	VSR10/VSR10	VSR12/VSR12	VSR14/VSR14	VSR16/VSR16	VSR18VSR18
Advance	4 Degrees	4 Degrees	4 Degrees		4 Degrees
Duration @ Lash/0.050 – In.	261/231	267/237	273/243	279/249	285/255
Duration @ Lash/0.050 – Ex.	261/231	267/237	273/243	279/249	285/255
Lobe Lift – In./Ex.	0.370	0.377	0.385	0.373	0.400
Rec. Rocker Ratio – In./Ex.					
302	1.5/1.5	1.5/1.5	1.5/1.5	1.5/1.5	1.5/1.5
327	1.6/1.6	1.6/1.6	1.6/1.6	1.6/1.6	1.6/1.6
350	1.6/1.6	1.6/1.6	1.6/1.6	1.6/1.6	1.6/1.6
383 (3.75-inch Stroker)	1.65/1.65	1.65/1.65	1.65/1.65	1.65/1.65	1.65/1.65
400	1.65/1.65	1.65/1.65	1.65/1.65	1.65/1.65	1.65/1.65
434 (4-inch Stroker)	1.65/1.65	1.65/1.65	1.65/1.65	1.65/1.65	1.65/1.65
V/Lift w/ Rec. RR (1.5/1.6/1.65)					
Valve Lift – In.	0.555/0.592/0.610	0.566/0.604/0.622	0.578/0.616/0.635	0.585/0.624/0.643	0.600/0.640/0.660
Valve Lift – Ex.	0.555/0.592/0.610	0.566/0.604/0.622	0.578/0.616/0.635	0.559/0.597/0.643	0.600/0.640/0.660
Recommended Spring #	74620LUN	74620LUN	74620LUN	74620LUN	74620LUN
Retainer (Steel)	75502LUN	75502LUN	75502LUN	75502LUN	75502LUN
Locks	77111LUN	77111LUN	77111LUN	77111LUN	77111LUN
(Pre-87 blocks) Rec. Lifter #	72400LUN	72400LUN	72400LUN	72400LUN	72400LUN
(Post-87 blocks) Rec. Lifter #	72403LUN	72403LUN	72403LUN	72403LUN	72403LUN

Notes: **Engines with strokes greater than 3.75 may need a reduced base circle cam for rod clearance (add –R to part number).** Post-1987 factory hydraulic roller blocks need the appropriate cam snout to accommodate the thrust plate. Be sure to state post-1987 roller block when ordering. The timing chain cam gear also needs to suit this later cam. These cams are intended for engines with intake valves between 2.02 and 2.08 inches diameter. If 1.65 aluminum rockers exceed budget, they can be replaced with lower priced 1.6 rockers. A reduction of 5 to 7 hp can result. On

INDUCTION SYSTEMS

Along with heads and cams, understanding carburetion and intake manifolds is among the most important aspects toward achieving high output. Squeezing all you need to know into one chapter isn't possible and this forces me to make statements I don't have the room to back up with lengthy explanations.

Understanding carburetion and induction does not give you an edge; it gives you an advantage. One year I dominated a particular professional factory class of racing in Europe to the tune of about 1 second per lap simply because I understood more about fuel atomization than my factory-entered opposition.

In this chapter we cover three aspects of the induction system: The air source point (i.e., hood scoops), filter cases, and filters. Next is carbs and calibration, and last, intake manifolds. To make the most power and torque we must optimize the ability of these components to deliver the most accurately calibrated, homogeneous, and coolest charge to the cylinder.

Sourcing and Filtering the Air

The induction system needs to supply the carb with cool, clean air in quantities sufficient to meet the engine's demand. The first step is to pick up cool air. If the

I threw in this shot of the carb for my big-block Chevy just because it's real eye candy. Sure it was far from cheap, but AED also produces some highly functional carbs at the other end of the scale. These custom-built carbs can be expected to run well with your combination right out of the box yet cost no more than regular carbs bought at your local speed shop.

air temperature is reduced from 150 degrees F to 80, as much as a 9.5-percent increase in "installed" output can result. If your engine was or will be dyno tested, remember that on the dyno it's being fed cool air so the increase we're talking about isn't power that's been gained as much as power that's not lost. On my Chevy-powered Pontiac Trans Am, changing from an under-hood air source of 190 degrees F to externally sourced air of 100 (by means

of an operative hood shaker intake) was worth 25 ft-lbs everywhere in the RPM range. At 5,500 rpm, this equates to 26 hp, so make induction the first rule for the sourcing of cool air. Fig 8-1 shows the benefits of both ram and cool air. Note that cool air boosts torque regardless of speed.

Supplying the motor with cool air often means the use of some kind of hood scoop. This isn't always convenient for many vehicle applications. An equally

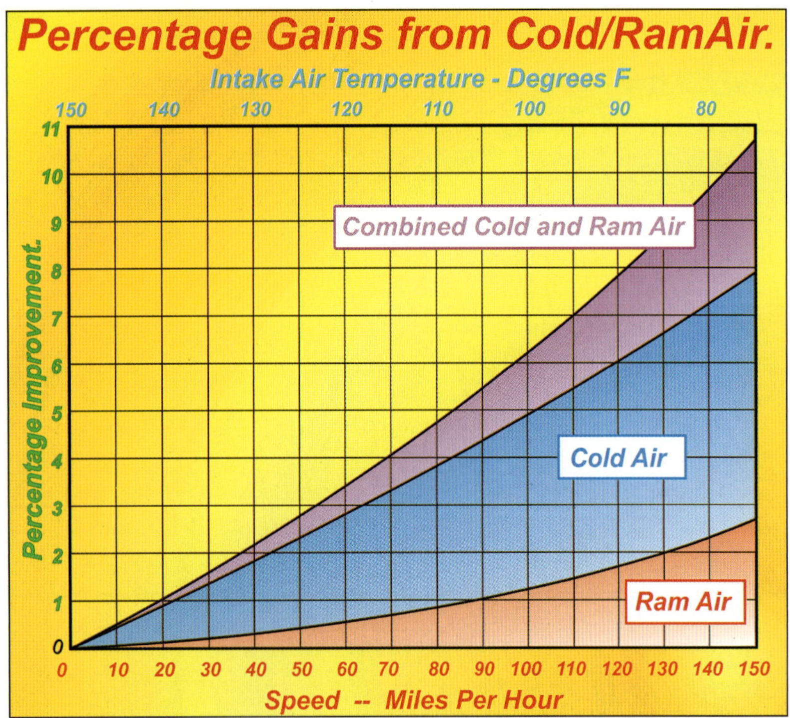

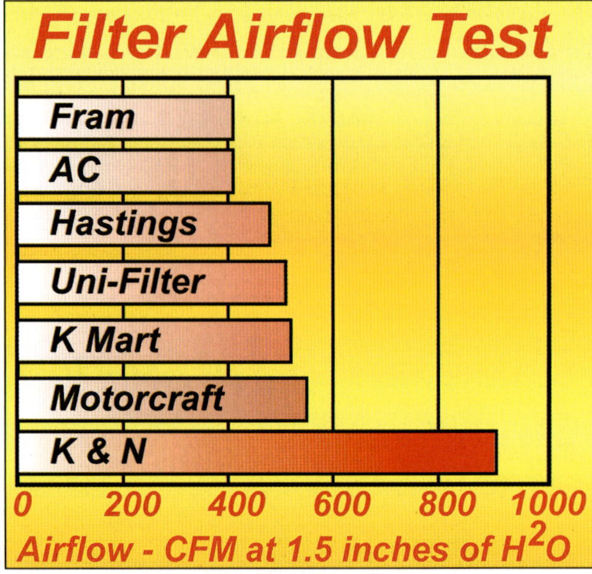

Fig 8-1. From this graph you can see what ram-air is worth compared to thermally managed air, i.e., cold air. It costs little to do, except time, and it's worth power all gained by virtue of extra ft-lbs of torque.

Fig 8-2. Here is how a K&N filter stacks up against some of the opposition. All these units were the same size. K&N research facility is equipped with high-tech equipment to optimize dirt arrest and retention so it is hardly surprising that these filters also show well in this area. For desert racing, these are my number-1 choice because I like to keep expensive engines from wearing out!

effective alternative (and ideal for a street driver) is to use one of Air Inlet Systems' (Ramairboxes.com) filter cases. This is designed to allow large-diameter heater trunking to couple the case to a cool air source at the front of the vehicle. These suggestions cost money, so why don't we just do without the filter and case, as is often done with drag-race engines?

First, dirt-laden air eats engines and that costs money. Running without a filter means saving a little now and spending a lot later. Using a filter means we pay some now and save later. The wear prevention aspect of a filter is obvious, but what's not is that using one will—even discounting the wear factor—make your car faster. This is an aspect of a filter that the majority of pro drag racers have yet to pick up on.

Assuming the filter has such a large-flow capability that this isn't a negative issue, here's how the filter helps performance. When engines are dyno tested they don't go anywhere, but when installed they do, and are subjected to high-speed airflow over and around the carburetor. If the carb booster action on the dyno is to be duplicated on the track, rule one is that the air supply to the carb must be steadied. A 100-mph airstream over the carb causes booster buffeting, and the signal it generates bounces around to the point of causing mixture fluctuations as great as 30 percent. There isn't a lot of sense to finely tune the carb on the dyno if in use it delivers random and rapidly changing mixture errors. Rule two is to never run your performance motor without an air filter.

I've spent lots of time testing air filters and have tested most of them. When it comes to selection, the bigger the filter the better. If a filter is too small, airflow will be restricted. A big paper filter of around 14-inches diameter and 4-inches deep will get the job done on engines up to 700 hp without power loss. Unfortunately, such filters need to be replaced at short intervals if full power is to be retained.

As a result of my tests as shown in Fig 8-2, I highly recommend the use of K&N filters because they have a much higher flow capability than any paper or foam material I've flow tested. In addition to high flow, they also have a high filtration capability and can hold several-thousand-percent more dirt before the flow is affected. This filter (which is used by Rolls-Royce) should be considered the first choice for a performance application. Although they're more expensive than a paper filter, they're reusable and do carry an unprecedented million-mile warranty.

Now consider the size of filter required to get the job done without incurring loss of flow. As a rough rule, 1 square inch of a K&N filter's folded pleat area (as opposed to unfolded) will supply the air requirements for 6 to 7 hp. This rating is good for road or drag racing and street use, but if the motor is intended for off-road use, the filter has to hold more dirt, so a rating of 4- to 4.5-hp per square inch is recommended. For a more precise filter size selection, get a K&N catalog. It has the sizing formulas and data a high-performance engine builder will need.

Carburetors

Before considering what parts to use, it's a good idea to establish what the functions of the carburetor are. Without assigning any particular priorities, a carb must satisfactorily perform three principle functions: satisfy the engine's airflow demand; deliver the required air/fuel ratio within close limits; and, deliver a mixture of adequate quality, i.e., suitably well atomized.

Failure in any area severely compromises output. For instance, a main jet that delivers 8-percent-less fuel at one point in the RPM range can cost 12 to 16 ft-lbs of torque. If that happens to occur at 6,500 rpm, almost 20 hp has been lost. I've seen poor fuel atomization cost an engine as much as 40 ft-lbs at low speed and inadequate airflow 20 hp at high speed.

The Q-Jet

If you're starting your project with a complete engine, you already may have a 4-barrel carb to work with—the stock Q-Jet. Stock Q-Jets are factory rated at either 750 cfm (1$\frac{3}{32}$ primaries) or 800 cfm (1$\frac{7}{32}$ primaries). However, because the factory sells carburetors based on application, these ratings seem arbitrary. I've flow tested so-called 800-cfm Q-Jets that returned better than 830 cfm.

With its small primaries, the Q-Jet delivers good street drivability and mileage, as well as power with its fairly substantial airflow. On motors up to 300 hp, there's little to gain from an increase in airflow.

However, motors with the potential to exceed 350 hp are easy to build, and the advantages of increased airflow will pay off. In this area, the Q-Jet potential goes largely unrecognized. Even if limited to the most practical modifications, a Q-Jet can be made to deliver a little over 1000 cfm. With more exotic modifications, such as installing bigger butterflies, this can be bumped to nearly 1,100 cfm. It can be achieved without sacrificing low end because of this carb's method of "on-demand-only" supply of air to the engine. To achieve this amount of airflow, refer to the Q-Jet chapter of my book, *How to Build Horsepower: Carburetors & Intake Manifolds* (CarTech).

It is easy to perform modifications to improve Q-Jet airflow. With these modifications come an increase in booster signal which counters the possible effect of over-carbureting the engine. A look into the primary side reveals that the butterfly shaft has excessive bulk and constricts flow. Thinning the shaft delivers about 10 to 12 additional cfm along with a big increase in booster signal. Working on the primary side to thin the bulk of the booster cluster also helps deliver more of what's needed. Diligent work here can return another 15 to 20 cfm.

On the secondary side, the butterfly shaft can be thinned and the butterfly knife-edged. This will add up to 6-cfm more, but the big gains come from setting the air-valve door at the optimum angle. This alone can deliver up to 70-cfm more. Unfortunately, such a move requires a flow bench and is outside of the scope of our activities.

For motors up to 375 hp, the fuel side of a Q-Jet requires little to be done past

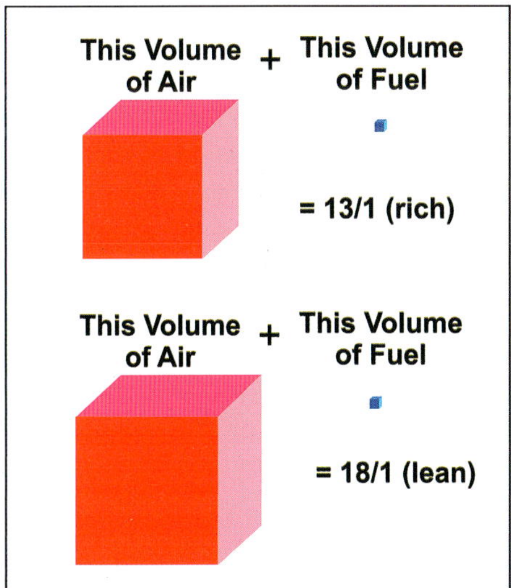

What you see here are the comparative sizes of cubes of air and fuel so that you can better appreciate what the carb has to deal with in terms of mixture delivery.

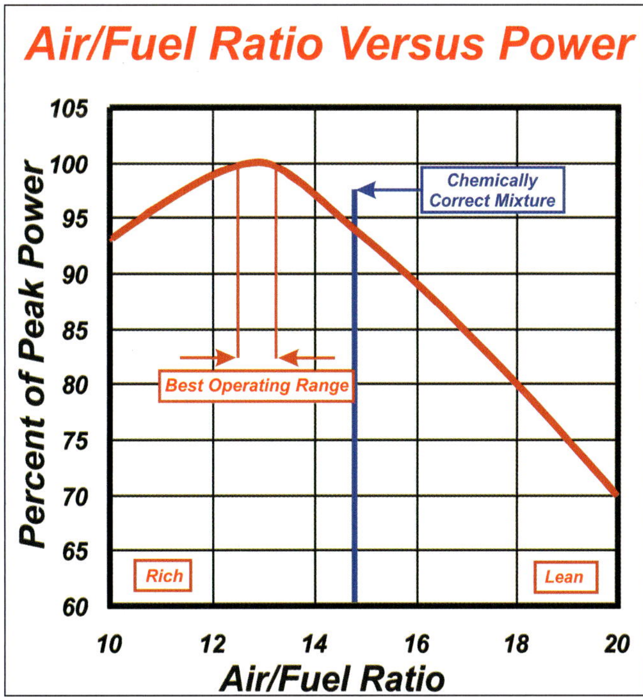

To find maximum power the mixture needs to be between 12.8:1 and 13.1:1. For best cruise economy with a conventional ignition system the mixture needs to be much leaner and a ratio of 16:1 to maybe 17:1 will be about best.

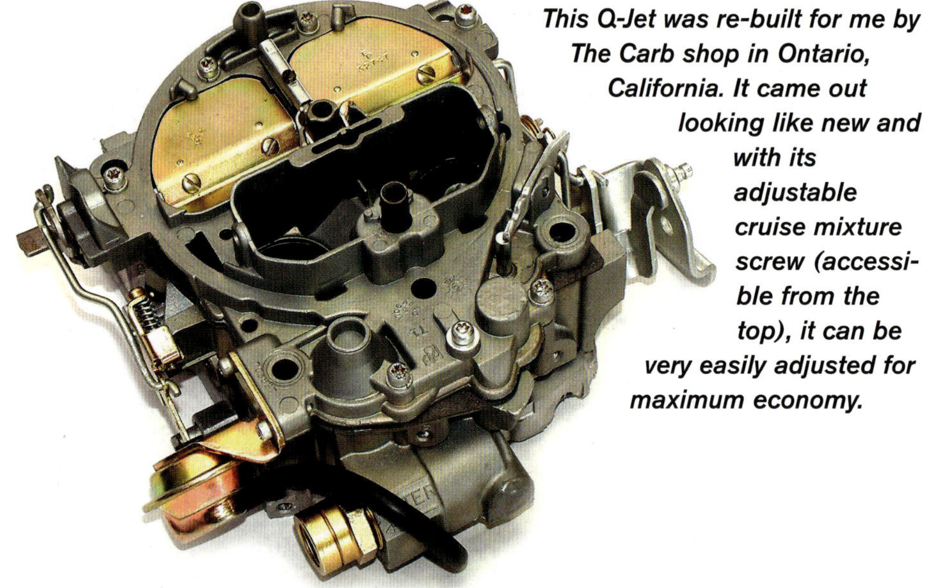

basic precautions to ensure adequate fuel flow. If your Q-Jet will be used on a high-output motor, there are additional factors to consider. First, do away with the fuel filter and use a large, high-flow inline filter. Check that the needle valve has adequate but not excessive flow capacity when fed with fuel at 4 to 6 psi. It should flow 1 pint per minute for every 90 hp.

If you don't feel inclined to check it out, the Carb Shop can supply a calibrated needle valve to flow the desired amount at the pressure used. To avoid vapor lock, insulate the carb from the intake manifold and shield it from unnecessary heat. Last, make sure the float bowl is equipped with the surge plate.

Calibrating the primaries is a relatively easy process because they need to be on the lean side for mileage. Once a needle close to requirements is installed, adjust the power piston up or down as necessary. If big cams are used, the calibration can become more complex. For outright power, the calibration of the secondary side is important, because some 75 percent of the air goes through this side.

Not only does the jet/needle combination used affect the mixture delivered by the secondary but also the degree to which the air valve is preloaded. The more resistance there is to opening this valve, the greater the pull on the fuel jets by the air valves' flow resistance. Also, the higher opening load restricts the air slightly.

These factors work in concert to influence the mixture delivered while the air valve is in the process of opening. A greater pre-load enriches the mixture slightly and a reduced pre-load leans it. While an excessive pre-load should be avoided, there's little to be gained from trying to run with the absolute minimum. As a good starting point, pre-load the air valve three-quarters of a turn and test drive the vehicle. If the engine goes slightly lean in the lower mid- to the upper mid-range, set the pre-load one-eighth of a turn tighter and retest. If it's necessary to use more than 1⅛ turns, go back to 3/4 of a turn and address the problem by needle and hanger changes. If the mixture looks okay to slightly rich, back off the pre-load by 1/8 of a turn. Half a turn should be considered the minimum pre-load.

Holley Carbs

The fact that Holley carbs have been around and used successfully for racing means they're in plentiful supply on the used market. However, without full knowledge of what and how to use and calibrate them, you may end up buying something unsuitable. For the majority of the engines built using this book there's a preferred group of Holleys.

Before getting into the subtypes of Holley, I want to make a point concerning the type of secondary operation you may choose. The 4-barrel Holleys are available in either mechanical or vacuum operation on the secondary barrels. Because low-speed driving isn't an issue, mechanical secondaries are used on all purpose-built race engines. This should in no way lead you to believe that vacuum secondaries won't perform and therefore have no place on your high-performance street motor. Mechanical secondaries are mostly a simplification for high-RPM-only applications.

Don't fall for the "monkey see, monkey do" trap. Just because something is used on race cars doesn't mean it's best for your street scorcher. If your vehicle is street driven, a vacuum secondary often will deliver a performance, as well as a mileage and drivability advantage, over a mechanical secondary carb.

There are two styles of Holley in the pool of carbs from which we can draw: spread bore, or square bore. Spread bores have two large and two small butterflies and are designed to be Q-Jet replacements. Square bore carbs, as the name suggests, have four evenly pitched barrels and the size differential, primary to secondary, is significantly less.

Spread bores were made in two styles, 650 and 800 cfm, and for the street can be made to work really well. All 650s are vacuum secondary and are well suited to street motors to about 300 to 350 horsepower, while 800s are generally mechanical secondary and are good on bigger-inch motors from about 325 hp and up. If the intake manifold to be used has a Q-Jet flange pattern—and almost all stock manifolds do—then you should put a spread bore Holley high on your priority list. Since carbs of differing flange patterns are adapted from one manifold type to another by means of an adapter plate,

For performance, the HP series of Holley carbs are a great choice. The most versatile of the group is the 750.

Interested in fuel mileage? If so, another option is the Edelbrock carb. Some tests conducted on my racer friend Tony Brown's motorhome tow vehicle showed a 17-percent improvement over the original carb. With that test under our belt, we are looking to do more mileage tests with this carb. Shown here are the 500-cfm economy version of the carb (top) and the Endurashine 800-cfm version we tested.

here's a point to remember: If power is a real issue, an 800-cfm street carb, such as a spread bore mounted on a two-plane spread-bore-style intake manifold, will significantly out-power a "race" 750 square bore installed on the same manifold via an adapter plate.

When it comes to carburetion on a two-plane manifold, the manifold (as we shall see later) effectively cuts the carb CFM seen by any particular cylinder almost in two. This means power production with a two-plane manifold is far more carb-CFM sensitive. A priority should be to increase carb airflow. That we will address later in this chapter.

Square Bore (4150) Carbs

Understanding the many nuances of carburetor design and function is more than just a casual performance edge if your goal is to go faster than the next guy, but what you need to learn takes up a book and then some. At this point it's best to avoid making common mistakes. The most common racer-mistake is to assume

that bigger cfm numbers equate with bigger horsepower numbers. The next most common mistake is to assume that a carburetor can have too many cfm for the engine. At first these two statements may appear to be in contradiction to each other. However, the problem isn't a question of too many carburetor CFM but too little booster signal to go with whatever cfm is available. To prove the point: I've run as much as 2,400 cfm of carburetion on a 350 intended to deliver economy and power right off idle. How did it work? Great, if 460+ ft-lbs. at 3,000 rpm (which was as low as the dyno would pull it) is anything to go by. What made this possible? Several factors, but the one I want to emphasize is booster signal. Given enough booster activity to ensure atomization, you can run as much cfm as the engine requires to meet its top end demand and still deliver the goods down low. That's in an ideal world. In the real world we have to deal with compromises, and the best compromise for what we are doing here is to make a selection heavily weighted

toward those Holley carbs that have a good cfm-to-booster-signal ratio. For all except high-RPM, wide-open-throttle race applications, a successful carb specification depends on making such a choice.

There are a number of carbs that will suit our purposes. The most versatile and goof-proof in terms of application and results are the 750-cfm models and a few of those with ratings either side. I center all my Holley carb selections to one degree or another around the 750; this has never failed to deliver results that, on occasion, are good enough to make the builders of fuel-injection envious.

If you're offered a deal on a 750 mechanical or vacuum secondary, take it because it can be suitably set up to deliver good results from engines of 325 hp to about 550. The 750s and many rated close to this have main venturis of 1.4-inches diameter. Within the 4150/4160 body, this is about as big as the main venturis can be made and still provide a signal of any consequence for a booster to perform at lower airflow rates.

This is an important issue if we are to select a carb with a wide working range. Without such a range carburetion will be poor at low main-venturi speeds, as is the case with straight or dog-leg booster 850s with their 1.5-inch venturis. A 1.4-inch venturi carb will work right down into the lower RPM range where many of the engines likely to be built out of these pages need to run, especially if it's of the vacuum secondary type. At main-venturi sizes greater than 1.4 inches, the signal drops by a sizable margin. This virtually mandates the need for a high-gain booster to amplify the otherwise feeble main-venturi signal if any sort of meaningful carburetion or atomization action is to be produced.

If you're buying your carb new and have determined what is needed, selection is a matter of ordering the appropriate part number from the Holley catalog. If you're buying used, there's the question of identifying the carb. First let's list what it is we're looking for. All the carbs that can work for us are listed in Fig 8-3. Avoid buying a unit that's been converted to alcohol unless it's something you specifically want.

Since the first edition of this book, I have found a carb company (AED Carbs) that is really good with Holley carbs. In essence, they custom build from scratch and ship you a carb calibrated for your engine, not one that is generically similar.

Fuel Atomization

If you don't want to be concerned with fuel atomization, sizing, or calibration, then get your Holley carb direct from AED. However, understanding what goes on won't hurt your knowledge base one bit so let us look at what goes in the fuel atomization department.

First let's categorize what's available from Holley in the order of preference. Most of my tests (especially boosters) were done with a 1.4-inch-diameter main venturi. An interesting point is that when these tests were run with a 1.5-inch main venturi, a 40-percent drop in booster signal was typically seen, while the bigger venturi only resulted in a 2-percent gain in airflow. Not a good trade off.

This leads me to my first recommendation. Don't use a stock 850 or any Holley with a 1.5-inch venturi unless it's used for an application of 4,000 rpm or more on a 400 or about 4,700 for a 350. If you feel the engine needs the airflow, an annular discharge booster with its higher gain will produce a signal on a par with a straight-leg booster in a 1.4-inch venturi. My recommendation is to use only these big-venturi carbs if an annular discharge booster is installed. Failing that, concentrate your efforts on the smaller-venturi carbs.

<table>
<tr><td>Preferred Holley Carbs</td><td>Fig 8-3</td></tr>
</table>

Street Use
Spread Bore: 650 and 800
Square Bore: 650, 700, 725, 750, and 780
- Vacuum secondaries are recommended, as are annular discharge boosters
- Best all around for 350/383 is a 750 reworked to flow about 820 cfm

Competition Use
Spread Bore: 850
Square Bore: 700, 725, 750, 780
Any Square Bore with main venturis no larger than 1.4 inches
- Best all around for 350/383 is a 750 reworked to flow about 820 cfm, or an 850
- Avoid straight-leg boosters
- Use dogleg boosters only on all-out racing engines where power below 4,000 rpm is not a requirement
- An 850 with annular discharge boosters is the best choice for all-around use, either on small race engines or up to 400-ci street engines
- A 950 race carb (no choke horn) is great if you can find one, but will usually cost too much. Use it only on an engine where power below 3,000 rpm is not not a requirement

If your engine is running a relatively large cam that gives a somewhat low vacuum, the transition onto the main circuit may stumble. This is because the butterfly is too far open and is past most of the transition slot. The fix is to drill a small hole in the primary butterflies and then close down the butterfly opening. The size of the hole required could be anywhere from 1/16 to 1/8 inch. Start small and test to see if the problem is cured.

If your car leaves hard and then feels like it is lying down, it could be the fuel in the float bowl has moved away from the jets. Jet extensions as seen here will fix this problem.

Increasing Booster Signal and Airflow

There are a number of steps that can be taken to improve Holley carb airflow and atomization capabilities. Here's where a file and the die grinder come into their own once more. Start by removing the throttle shafts and file down the shafts to streamline them between the screw threads. Reassemble with the screws suitably shortened and be sure to put Loctite on the threads so your motor doesn't eat wayward throttle-shaft screws. Reworking the shafts typically increases booster signal and improves the airflow by about 25 to 30 cfm. This mod is well worth the effort for a street machine because there are no downsides. Low-end output is often marginally improved and top end can pick up to the tune of 8 to 12 hp on a motor in the 400-hp range. I've seen as much as seven horses on an otherwise 300-hp motor. This is especially effective when two-plane manifolds are used.

The next move can be achieved one of two ways. The first will cost, the second won't. It doesn't take the eye of experience to realize that the air's entry into a typical Holley has to work its way around some un-streamlined obstacles. The use of a K&N Stub Stack (yes, they really do work) fixes much of this and results in a flow and booster signal increase. The booster signal gain is usually small, but the flow increase isn't. On an 850 it typically amounts to 30 to 35 cfm, and 25 to 30 cfm on a 750. The Stub Stack allows the choke to be retained in a functioning form so this move is ideal for a street machine. If cold-starting convenience isn't an issue, your die grinder can be put to work streamlining the carb entry. These airflow mods are relatively elementary and can be applied across the board to virtually any Holley carb, whether it's a square or spread bore design.

Calibration

Regardless of how good it is, no carb will deliver if it's not accurately calibrated.

The correct way to calibrate a Holley's WOT mixture is via the power valve restriction channel. Usually they cannot be calibrated, but this performance Holley has jets to do just that. The main jets should be calibrated for best cruise mixture with the jets shown here sized for maximum power mixture.

This is AED's base entry-level carb. Don't let the fact that it is one of the least-expensive performance carbs available lull you into thinking it produces economy price results. My tests have shown that in eight cases out of ten it will out perform a carb costing at least 30-percent more, and when you are working on a budget that's a point to consider carefully.

Some famous professor once said (and I'm sure it was in jest) that a carburetor was a wonderfully ingenious device for giving the incorrect mixture at all engine speeds.

The mixture curve that is so often referred to but rarely explained is trimmed via these jets/bleeds shown here. Making the jets at the top bigger leans out the bottom of the RPM range. Making these same jets smaller richens up the low-speed mixture. Making the bottom jets bigger leans out the top end and making them smaller richens it up. On top of that we have to add the effect of the . . .

. . . air corrector jets, or air bleeds (brass jets either side of booster) as they are sometimes called. Making the air corrector jets for the main circuit bigger leans out the top end of the RPM range and smaller richens it.

Not true! Such a statement applies to a carb calibrated by someone lacking the necessary skills. The biggest problem concerning carb calibration is knowing what the mixture is at any one moment, closely followed by knowing what adjustment will fix it. Plug reading is really a dubious pastime at best. I know someone out there is saying, "If it's okay for the late Smokey Yunick, it's okay for me."

The mixture may well get really close to what's needed if Smokey does it, but can you afford to wait while you accumulate Smokey's years of experience? I doubt it. Plug reading will only get you into the ballpark, and then only if you are good at it. With the advent of oxygen sensors, a much better if more-costly option is now open to you. Although it will set you back a little over a \$100, an oxygen mixture analyzer from K&N, Edelbrock, MSD, or one or two other manufacturers is a good investment. Of course you can avoid calibrations woes if you take my advice and go to AED for your Holley.

Manifolds

We are going to look at two fundamental types of intake manifold here: the

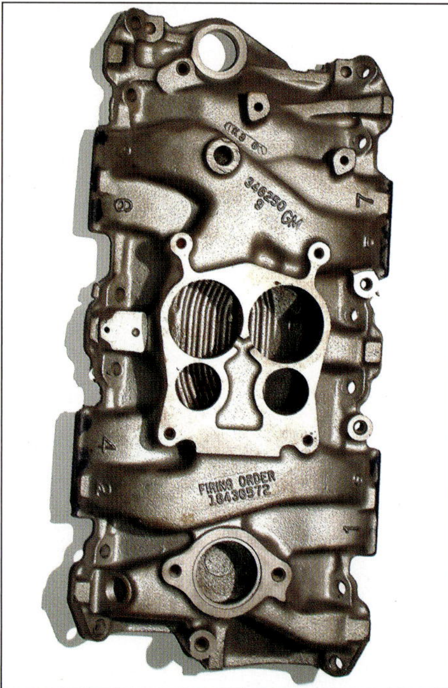

The class rules for my dirt car do not allow the runners to be ported. What I did to maximize output was to cut heat input into the intake charge. All that can be seen from this side is the gray-colored thermal barrier coating on the interior of the plenum/runners. Turning the intake over though reveals . . .

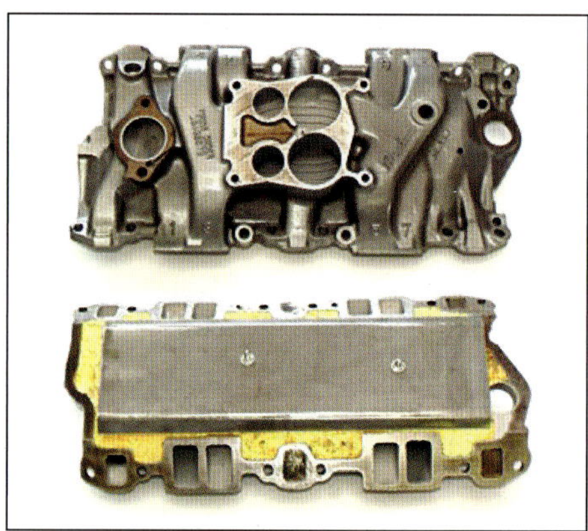

. . . the extensive work done to isolate the hot water from the rest of the manifold, and the insulation to stop hot oil from heating everything up. This intake was worth 20 ft-lbs everywhere in the RPM range. It worked so well that after each dyno pull the intake temperature dropped rather than showing the more normal increase.

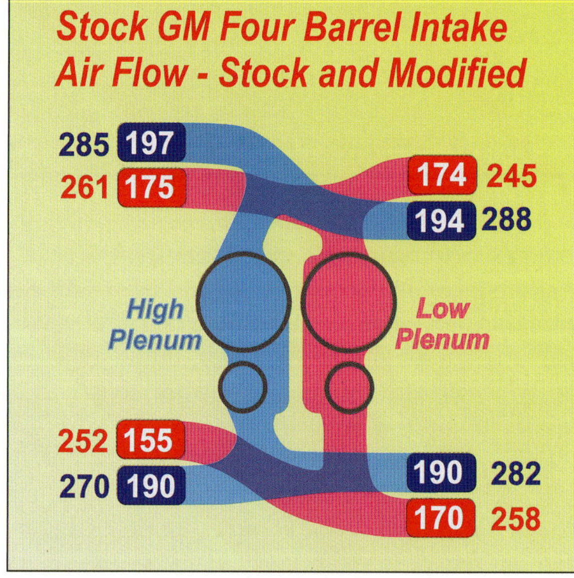

From this diagram you can see how the runners of a two-plane intake are laid out. With this pattern of runners, each half of the carb sees an induction pulse every 180 degrees (that's why it is also referred to as a 180-degree intake). The numbers in the runner openings are the flow figures for the stock intake. The adjacent numbers are for the flow after I spent about 30 hours porting it. The increased flow accounted for 26 hp on a motor starting with some 300 hp.

Here's the 820-cfm AED Holley that is one of my favorite pieces. In the three years I have had it this carb has never failed to produce top-notch results.

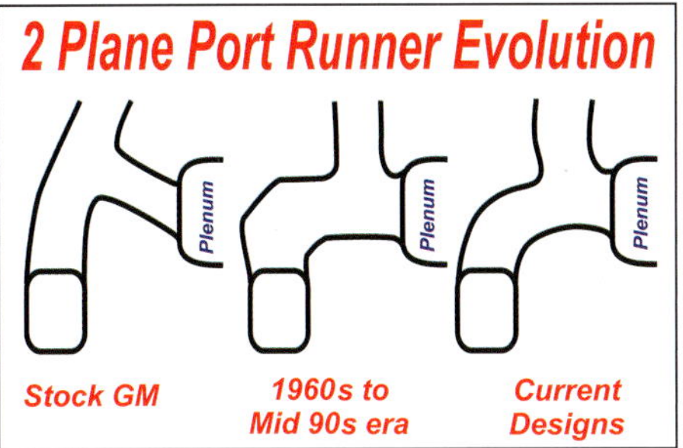

What you see here is the basic evolution of the port runner shape of two-plane Chevy intake manifolds. Do not bother with anything less than a current design!

Here are the four intake manifolds I dyno tested. They all produced top-notch results so there is little to choose in that respect. The least expensive was the Professional Products Crosswind manifold seen here at far left.

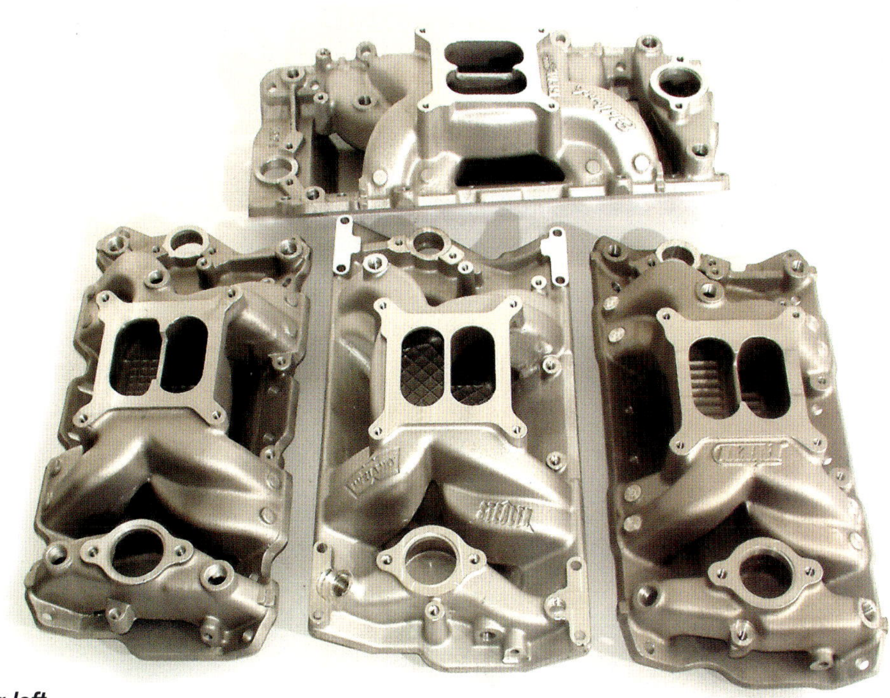

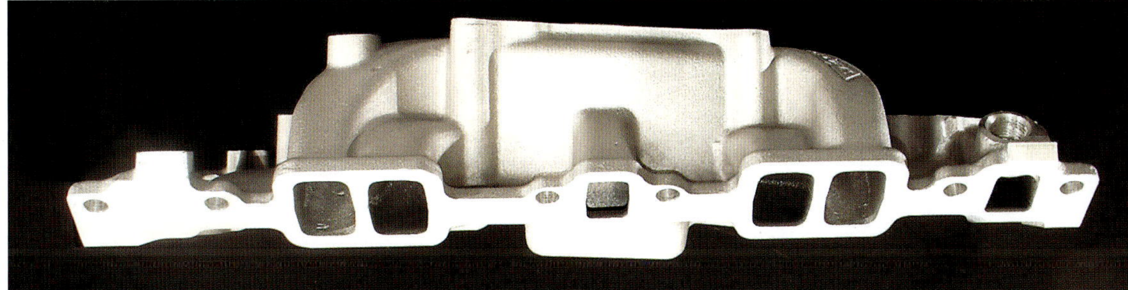

This is a regular Edelbrock Performer with the heat crossover passage. This intake is about an inch lower than most air-gap-style intakes. If hood clearance is an issue,

then this intake with the crossover blocked will produce results almost as good as an air-gap intake. If manifold heat is a must, due to harsh winter weather, this is the intake to use.

two plane, and the single plane, both in single 4-barrel carb configuration. Check out the nearby drawing to understand exactly what a two-plane intake is.

All stock intakes are of this configuration and all have exhaust crossover to heat the intake charge. The question here is do we need manifold heat? If you live in the southern part of the country, especially the Southwest, it probably can be done away with. If you have an annular discharge carb, I strongly advise running a cool manifold, because it's worth a sizable number of ft-lbs even on a mundane motor. A typical gain as a result of blocking the intake heat is 10 to 11 ft-lbs. If a cooler intake is used, about three-quarters

This is what the runner shapes of an Edelbrock Performer RPM look like. As you can see, they are pretty streamlined and, compared to earlier designs, produce good flow as well as strong port velocity.

Here is one of my street screamer 350s with an AED carb and a port/plenum-matched Endurashine Victor Jr intake. This 10.5:1 pump gas engine made 548 hp. It would drive at 2,000 rpm and would turn up to 8,000 rpm at the top end.

of a compression-ratio-increase above the previous limit is possible, and that's worth a like amount.

The downside of eliminating intake heat is minimized if the carb is equipped with annular discharge boosters, and this is the type of booster I recommend. Aluminum manifolds pick up heat and are less likely to need additional heat from an exhaust crossover. If you intend to use a two-plane as opposed to a single-plane manifold, remember that carb CFM becomes more important. Short changing the carb by 100 cfm could cost 40 hp at the top end while gaining nothing at the bottom end. I regularly use 820- to 950-cfm carbs on street-driven two-plane equipped engines of 350 to 383 inches.

Cost-Effective Aftermarket Intake Manifolds

The stock four-barrel manifold can be modified with a die grinder to good effect. Some basic porting mods performed on a stock iron manifold will have an effect on both airflow and power, but it is time consuming. There are good reasons for swapping out a stock iron intake for a good two-plane aftermarket one. Not the least of these is a big increase in horsepower, a sizable increase in torque, and a big weight saving off the front of your vehicle. The

This is Wieand's (division of Holley) air-gap-style intake. Because it does not have a cut-away divider wall, which allows some communication of left to right plenums, it is more sensitive to carb CFM. This manifold likes the bigger cfm numbers to perform to its full extent.

At the time of this writing I have been having great success on engines capable of over 500 hp with the Super Victor intake. Here is one that I have just finished prepping. Note the plenum clean up and the smooth but non-polished internal surface.

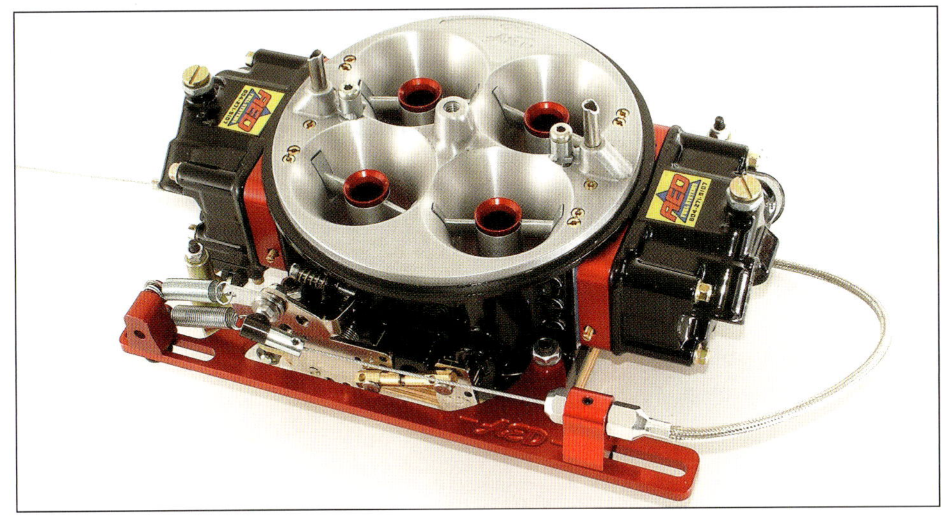

Just some exotica to make a point and for you to drool over. The AED 1250-cfm Dominator-style carb shown here was claimed to be about as drivable as a Holley could be made. The boss at AED claimed it almost made a nonsense of having a carb "too big." Just to see how that worked out, I ran a test on the 355 shown nearby. With a suitable intake such as . . .

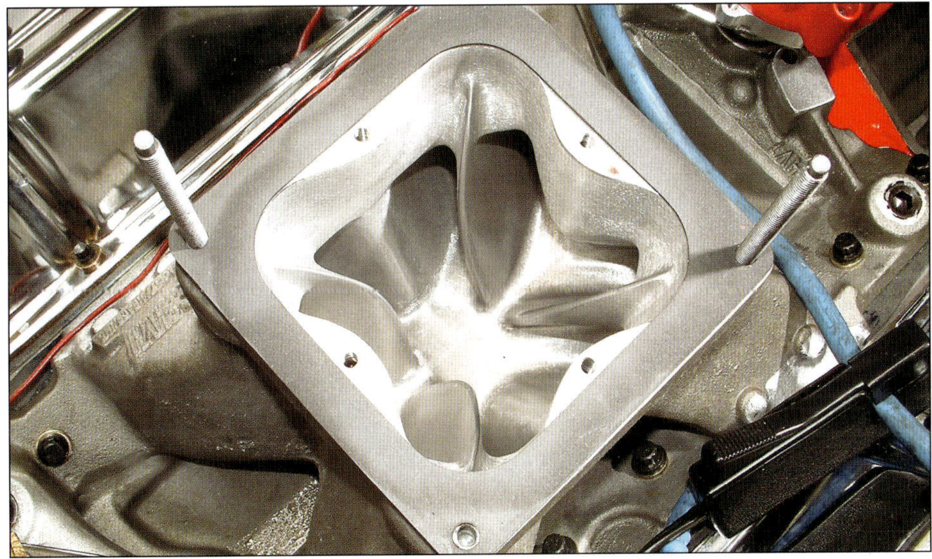

. . . you see here (this example is an Edlebrock manifold but Dart also make one like this). It produced 584 hp at 7,000 rpm. It also ran perfectly at 2,000 rpm and only gave away 8 ft-lbs to a 750 carb at this low RPM. The top end increase with this carb and manifold was well over 20 hp. It also ran flawlessly to over 8,100 rpm where it still made 564 hp. Now if AED can make a monster carb like this work both at the top and bottom, you ought to be able to get some idea what they can do with a more normal-sized 4150-style carb such as 750.

port development and thermal management. The nearby drawing shows how the port shapes have evolved over the years. Only buy the latest forms if you are buying used (they have been around since the 1990s and there are plentiful used ones).

So how well does the current crop of high-tech two-plane intakes do? Basically you can figure that on a 350 that nominally makes, say, 325 hp on the stock intake, you can be looking at 355 when the intake is swapped out for a current high-performance air-gap-style intake. Make that a 383 and the power increase will be bigger yet. As for top-end capability, I have seen (as of 2010) almost 570 hp with a two-plane air-gap intake on a street driver small-block Chevy.

Most of these modern two-plane intakes are available with cross-over heat and no under-runner air-gap, or in air-gap form. The key issue here is the exhaust heat. Deleting this is worth 5 to 10 hp, depending on the spec of the rest of the engine. However the difference between a heat deleted non-air-gap and an air-gap intake is relatively small—typically 3 to 5 hp. If you live in northern climates and have to use your street rod for daily commuting, then an exhaust-heated or better yet a water-heated intake are the way to go. To get water heating to the intake you will need to do a few hours of modifying but the results are worth the effort.

The examples shown in the nearby photo were all of the air-gap variety and were all dyno tested by me. They all proved highly functional and the results were so close as to be indistinguishable except to maybe a couple of thousandths-of-a-second max at the drag strip. As far as cost goes, the least expensive by far was the Professional Products Crosswind intake, thus making it the most cost-effective of those tested. It was also with a 930 AED carb, one I saw 539 hp on!

question is: which manifolds are likely to give you the best return for your money? In terms of function, the modern aftermarket two-plane has evolved into a highly effective piece of speed equipment for those of us building naturally aspirated, street driven small-block Chevys. The two key issues have been

EXHAUST SYSTEMS

Enthusiasts often see putting together a functional exhaust system as black art. Complex though they can be, it's not black art. Fortunately for us the engine we're dealing with has been under development as a high-performance unit for more than 55 years. This means much of our work in terms of putting together a functional exhaust system already has been done for us. This factor alone means that we don't necessarily have to design our own components. All that's required is to find suitable selections from what's financially viable and then put such parts together in a functional manner.

As it happens, there are plenty of functional—though not always affordable—exhaust system components on the market, as well as a few that have no business on a performance machine. As a result, one could be forgiven for concluding that assembling a working exhaust system would be easy. If what I so often see used in practice is anything to go by, this proves to be far from the case. In round figures, I estimate that fully 75 percent of the exhaust systems I see are far from optimal because either substandard components are used or the system is poorly spec'd. Either way, a great deal of power can be lost. Before going into design detail and parts selection, it's a good idea to

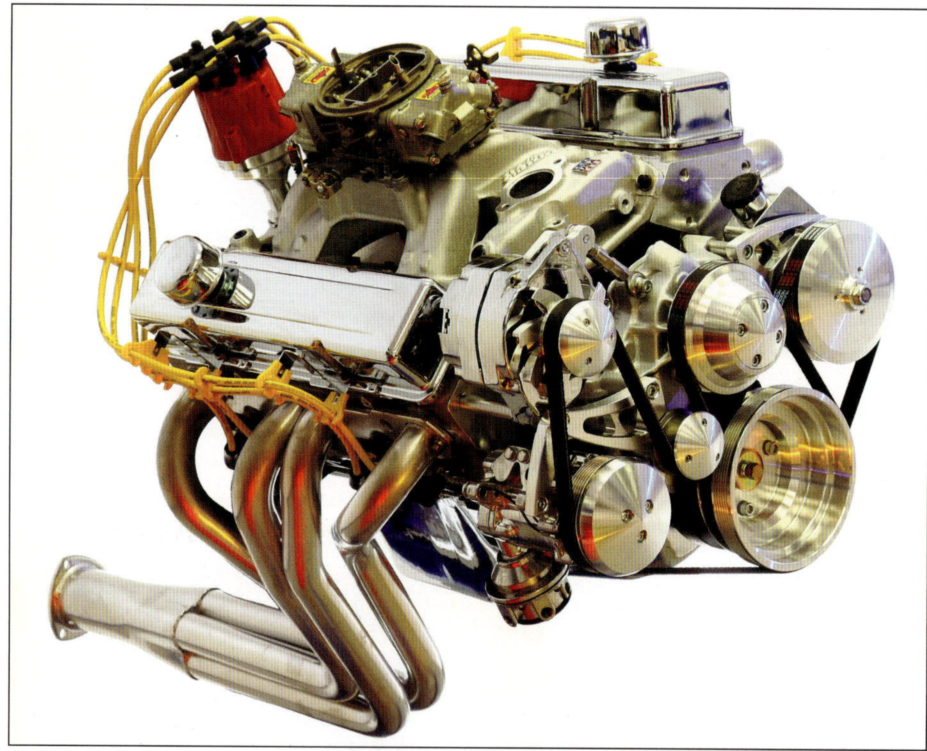

This engine build may look pricey but that is an illusion brought about by the billet March belt drive system and the polished accessories. Take away these sharp-looking accessories and you are left with a 500-plus-hp 383 that falls well within the budget constraints of this book.

explain the importance of a functional exhaust system. This will allow you to set realistic goals and improve your ability to choose functional components and to install them in a more optimal fashion.

Exhaust Initiated Induction

Engines such as the small-block Chevy are regarded as 4-cycle engines. While this rather obvious statement may

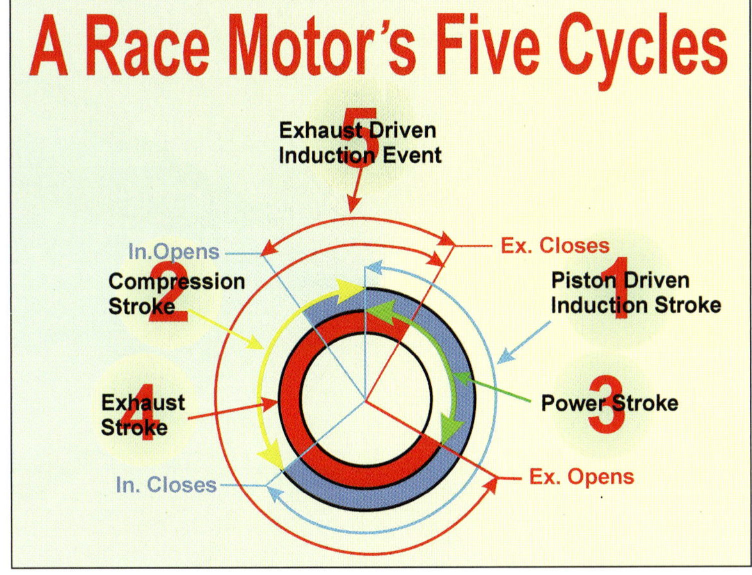

Starting at cycle No. 5, we see that the exhaust-generated vacuum starts the intake charge moving into the cylinder way before the piston even starts to go down the bore. As the crank rotates farther we get to cycle No. 1. This is what is normally considered the induction stroke that draws all the new charge in. In an ideal situation, cycle No. 5 has cleared the combustion chamber and put a considerable amount of kinetic energy into the incoming charge before the piston even starts on its way down the bore. The result is an engine that can achieve a volumetric efficiency well over 100 percent. The bottom line is a good exhaust system is worth a lot of extra torque, horsepower, and (best of all) extra mileage.

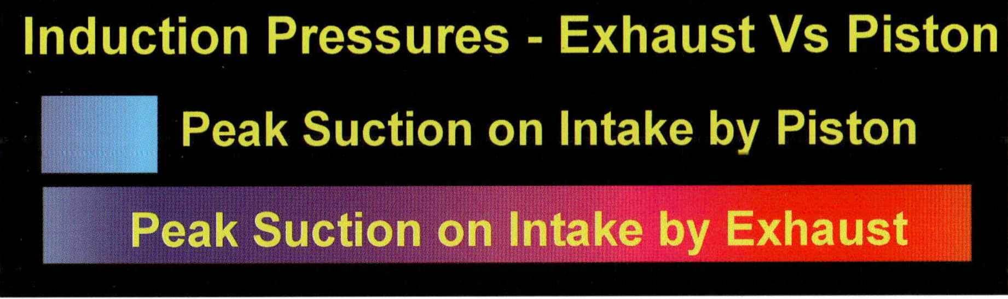

Here is a comparison of the induction suction created by the piston going down the bore and that created by the exhaust tuning. As you can see, the exhaust is the big player here—not the piston motion down the bore.

be so for a regular street powerplant, it's by no means totally true of a race engine.

If we consider a well-developed race engine in terms of the normal, induction, compression, expansion (power stroke), and exhaust cycles, we have a fifth element added. If the exhaust system is suitably length-tuned, negative pressure waves can scavenge the combustion chamber during the valve overlap period that exists at the end of the exhaust stroke and beginning of the intake. To understand how this increases an engine's

breathing, let's consider the cylinder displacement and combustion chamber volumes of a typical high-performance 350.

In traveling down the bore, the piston of a 350 displaces 727 cc. If the engine has a compression ratio of 12:1, the total combustion chamber volume above the 727 cc will be 63 cc. If a negative pressure wave draws out the residual exhaust gases remaining in the combustion chamber at TDC, then the cylinder can draw in 790 cc (727 + 63). The result is that this engine now runs like a 385-ci

instead of a 350. But there's more to this than just scavenging the chamber. If enough energy is put into the incoming charge by the exhaust, then it's possible to cause the cylinder to fill to a pressure greater than atmospheric at the time of intake valve closure.

Compared with the intake, exhaust tuning is far more potent and can operate over as much as 10 times the RPM band. To put the importance and capability of the exhaust pressure wave into perspective, let's consider a few numbers. As we know, air moves from one point to another by virtue of a pressure difference. The charge-inducing pressure difference (suction) is normally associated with the piston traveling down the bore on the intake stroke. The better the head flows, the less suction it takes to fill (or nearly fill) the cylinder. For a highly developed race engine, the pressure difference between the intake port and the cylinder shouldn't exceed about 10 to 15 inches of water (about 0.5 psi). If it's any higher than this, it demonstrates an inadequate head is being used. For a budget race motor such as we're dealing with, about 20 to 25 inches of water (about 1 psi) is about the limit if decent power is to be achieved. From this we can say that, at most, the piston traveling down the bore exerts a suction of 1 psi on the intake port. On an unrestricted NASCAR Cup Car motor, the finely tuned exhaust will exert a suction of 6 to 7 psi on the cylinder. Because this occurs during the overlap period, much of this suction is applied via the open intake valve to the intake port.

You can see that the exhaust system can draw on the intake port as much as 500-percent harder than the piston going down the bore. Under these circumstances, it's the exhaust that's the primary element of induction, not the piston traveling down the bore. With such a system, the charge in the intake port can be traveling into the cylinder at 100 feet/second even though the piston is still parked at TDC! In practice, the

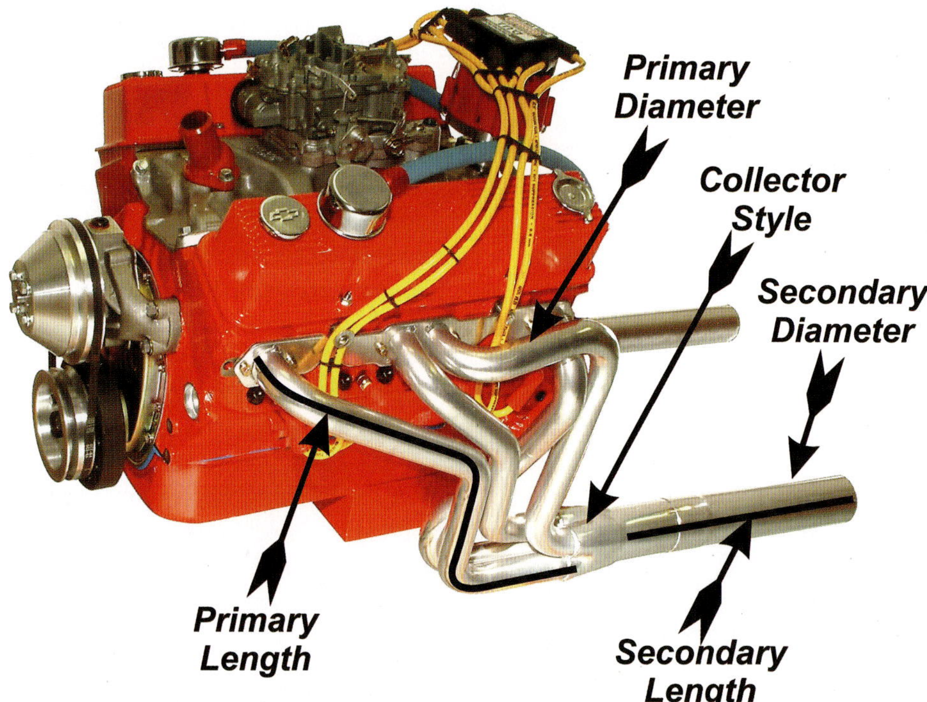

This is my dirt car engine. Considerable time was spent fine-tuning the parameters shown here. These subjects are covered in this chapter.

exhaust phenomena turns a race engine into a 5-cycle engine with two consecutive induction events.

Now that the vital role the exhaust system plays toward the development of power has been highlighted, it will be easy to appreciate that learning to tune the system (with or without mufflers) will be a winning factor. Here the emphasis is on race engines, and this implies the use of big cams with an overlap period sufficient to allow the through flow of gases around the TDC overlap period. As the overlap period is reduced, so is the opportunity to achieve this. By the time the overlap is down to about 50 degrees, the ability to scavenge the chamber and impart energy to the intake charge to assist filling is minimal. However, that is certainly no reason why a hot street motor shouldn't benefit greatly from exhaust tuning. Unfortunately, in all but a few isolated instances, the system "tuning" is disrupted by the presence of a usually ill-conceived muffler system. Learning how to tune systems with mufflers will be an important skill, because an increasing number of sanctioning bodies and racetracks are mandating maximum noise levels.

At the time of this writing, it seems that getting power and reducing noise are, for the most part, mutually exclusive. Historically this has been so, but building a quiet system that allows the engine to still develop within 1 percent of its full, open-pipe power is entirely practical. In this chapter we'll look at system requirements to achieve this desirable state of affairs. Knowing what it takes in this department can easily endow your engine with a 20- to 40-hp advantage over your less-informed competition. Read on and take note.

Low Budget and Basic Iron

The exhaust system actually starts at the exhaust valve and continues through the port and on into the exhaust system. We dealt with exhaust ports in the cylinder head chapter, so our investigation will start at the exhaust manifold face.

If the cost of a set of headers for your particular installation looks to be outside your budget, you'll have to make the best of a set of iron exhaust manifolds. This isn't as bad a situation as it may first appear, especially when a street cam is used. Certain classes of oval track racing call for the use of stock iron manifolds. As has been so well demonstrated by Randy Brzezinski at Brzezinski Racing, these can be modified to good effect. Brzezinski has spent a considerable time developing these "stock" iron-type manifolds and has achieved power gains equal to about two-thirds of that produced by a good header system.

However, the modifications entailed are often more than just grinding the manifolds to make them flow better. In many cases, the manifolds have deflector plates welded in to cut cylinder-to-cylinder interference. Replicating Brzezinski's best efforts could well be as costly as buying one of the cheaper brands of headers. But there's a lesson to be learned here. Randy Brzezinski informed me that a fair proportion of the gains he sees are achieved by porting the manifolds in much the same way we port the heads.

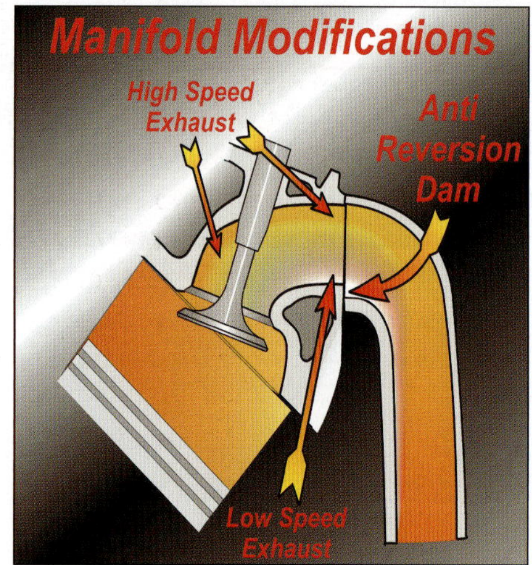

If you are porting a set of iron manifolds, match up the top and sides but not the bottom. A step here acts as an anti-reversion dam to cut the reverse flow on the floor of the typical Chevy port. When this reverse flow occurs it brings about a substantial reduction (20 to 30 ft-lbs) in low-speed torque.

Having proven the effectiveness of Brzezinski's manifolds on my own dyno, I made a point of doing basic porting on several sets of iron manifolds of differing configurations to see what they'd typically do. I was surprised at the results. On engines in the 280- to 350-hp range, it was common to pick up at least 10 hp, with 15 being about the norm—on one occasion I saw as much as 22 hp. I have to stress at this point that the porting done was only basic. It involved port matching to the manifold gasket (usually a Fel-Pro) at the top and sides, but the bottom mismatched header style. This mismatch acts as an anti-reversion dam to the slower-moving gases at the bottom of the port. By imped-ing their reversion, low-speed torque is increased without loss of high-end out-put. When grinding the manifold ports, attention was paid to helping the exhaust around the usually tight turns immedi-ately adjacent to the manifold face. With regular porting tools it's not possible to access the port far in from the manifold face, but the area we can get to is the most critical. The other area that typi-cally needs attention is the point of exit into the down pipe. Here most manifolds seem to be restrictive even for small blocks with as little as 150 hp. Enlarging this area too much will obviously cause the manifold to fail. The technique to apply is to smooth out irregularities as far as possible on the approach to the exit and enlarge the exit diameter by about 1/8 inch or so. From here, plumb in a 2½-inch down tube.

Pipe Diameters

Is bigger better? The answer to that most certainly is no. For a performance street motor, having a wide power band produces almost as much driving pleasure as having a lot of horsepower. The best out-put at any particular RPM is seen when a certain exhaust velocity exists. This means the best-sized pipe for 3,000 rpm will be

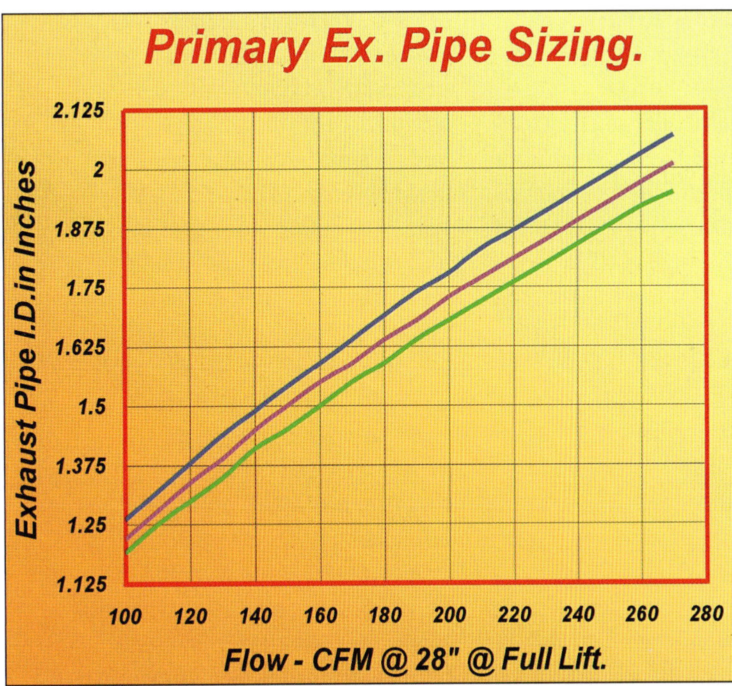

Fig 9-1. This graph can be used to make a near-optimal primary pipe size selection first time around. First, you will need the exhaust port flow for your heads at the valves full lift point. Next, deter-mine which of the graph lines fits your needs. The green curve is for best all-around results on the street. The purple line is more for street/strip applications where low-speed torque will suffer margin-ally but the drag strip results will look good. The blue line is for high RPM race operation where the lowest RPM seen is about 75 to 80 percent or more of the maximum.

An example pipe selection works like this: the heads have 175 cfm at the valve lift used and the application is street/strip so we choose the purple line. Find the 175-cfm point along the bottom scale. Next follow this up to the purple line then directly across to the scale on the left hand side. This indicates a primary pipe size of 1.625 ID is needed for the job.

different to what's optimal at 5,000, so some compromises must be struck.

The flange bolt pattern on stock small-block Chevy heads makes it diffi-cult to install headers with pipes larger than 1¾ inches in diameter. Above this diameter it's necessary to use a Hooker adapter flange, and this, together with the more expensive header, starts to push cost out of the budget category.

The industry standard for pipe sizes for the street-performance small-block has been established at 1⅝ inch. This not only is easiest to produce, but also works well on motors from 200 to 375 hp, and so covers about 80 percent of require-ments; 1¾-inch headers are made for pop-ular applications, but I only advise their use if your power target is above 375 hp.

If you've cut expenditures in other areas by doing the labor-intensive work yourself (such as head porting), then the power possible may outpace what a 1¾-diameter header can handle. If this is the case, use Fig 9-1 to determine the engine's requirements.

So much for primary pipe diameters. Let's now consider the collector/secondary pipe diameter. The first point is that the secondary diameter is as criti-cal as the primary. For good all-around street performance, a 2½-inch secondary works well. A 3-inch secondary is only of advantage when the horsepower is up past the 375 mark. If you have a really high output unit in mind, then the col-lector size should be 1¾ times the diame-ter of the primary pipes.

Headers and Low System Flow

From the effect that porting iron manifolds has, we can see the manifold's flow capability is important. If the manifold doesn't dispense with the exhaust quickly and effectively, the high-pressure blowdown period of one cylinder will affect the ease with which a connected cylinder clears exhaust on the exhaust stroke.

Pumping against a higher pressure increases pumping losses. Also, a high pressure at the port during the last phase of the exhaust stroke means the combustion chamber is at a higher pressure than the intake port. When this is the case, the piston will have to travel a short way down the bore before it can draw in any fresh charge. This means the higher combustion chamber pressure has the same effect as reducing the engine's cubic inches, and that's something we don't want. Installing headers improves the flow and separates ports so there's less interference. The effect headers have on a short-cammed street engine proves worthwhile even if the cam is short and the rest of the exhaust system has limited flow and produces backpressure. Fig 9-2 shows how much extra power can be seen from a typical, near-stock 350 with a full emissions-legal exhaust system having a high backpressure. Of course, the full potential of the headers isn't realized until the system's backpressure is reduced, and that's a subject we'll get to shortly.

Pipe Lengths

Misconceptions concerning exhaust pipe lengths are widespread. First let's take the often quoted phrase "equal-length headers." I've heard engine builders and racers make a big deal about their engines having headers with the primaries uniform to within 1/2 inch. This raises the question of whether or not they knew what was wanted with 1/2 inch! I doubt it, and though unlikely, the system could have all the

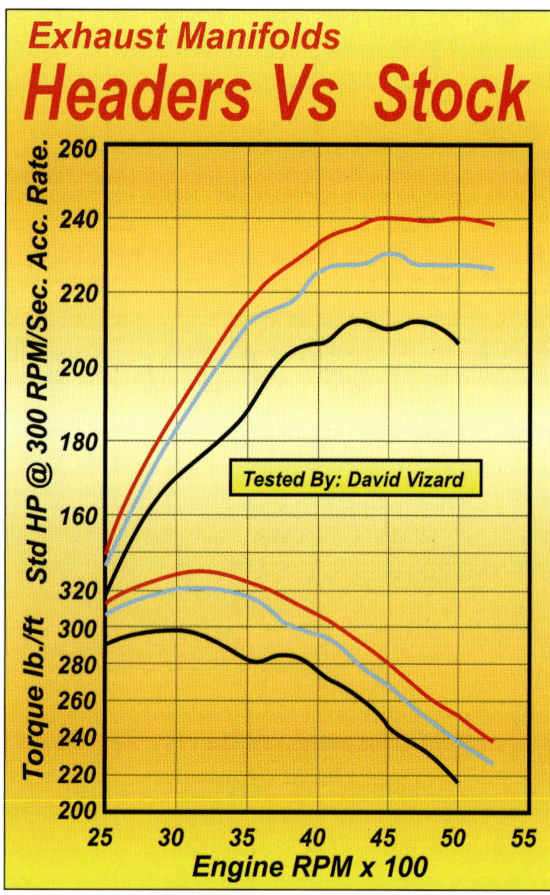

Fig 9-2. This dyno test shows that even if you are forced to use catalytic converters with less than adequate flow, headers still pay off. I ran this test run on a 1979 350 Z28 (yes they really were that bad) and it shows the power with the complete stock exhaust system on (black curves). Only 211 hp and 300 ft-lbs is hardly going to be enough to make a 3,400-pound car a performance car, but that was it for that era. Installing a set of 1⅝ headers (1½ ID) bumped the torque by 20 ft-lbs and the HP by 19 (blue curves). Installing high flow cats and mufflers generated the curves shown in red. This amounted to a total of 29 hp and 30 ft-lbs increase. All this came with almost a 2-mpg fuel saving.

pipes equally wrong within 1/2 inch. Building a set of headers to such precision is close to a waste of time for a race engine and totally ludicrous for a street engine. Under ideal conditions, it's entirely practical for an exhaust system to scavenge the cylinder at near maximum intensity over a 4,000-rpm bandwidth. Most race engines use an RPM bandwidth of 3,000 rpm or less. If the primary pipe scavenging effect overlaps this band, then it matters little that one pipe tunes as much as 1,000-rpm different from another. Since this is the case, pipe lengths varying by almost a foot have little effect on power.

A good street motor can have a working RPM range as high as 6,000 rpm. One way to spread that 4,000-rpm band is to use differing primary lengths. A positive, power-increasing attribute of differing primary lengths is that it allows larger-radius, higher-flowing bends and more convenient pipe routing to the collector in often confined engine bays.

Apart from the reasons just stated, there's another good reason why worrying about equal primary lengths is a waste of time. In practice, a two-plane cranked V-8 engine such as the small-block Chevy is simply insensitive to quite substantial primary length changes. The reasons for this may be less than obvious. My experience here indicates that inline four-cylinder engines appear more sensitive to primary pipe length, but a small-block Chevy isn't two inline fours lumped together to form a V-8. It is, in fact, actually two separate V-4s, and as such it doesn't have evenly spaced exhaust pulses down each bank. With a conventional header (as opposed to a 180-degree installation), exhaust pulses are spaced at 90-180-270-180-90, and so on. The discharge of two cylinders only 90-degrees apart appears at the collector

If you want your headers to look good and last, go for the ceramic coated ones as shown here. These days I don't use anything else.

as one large cylinder. This means the primaries act like they do on a four-cylinder engine, but the collector acts as if it were on a three-cylinder engine having different sized cylinders turning at fewer revs. (Isn't life complicated?) This, plus the varied spacing between the pulses, may be the cause of the system's insensitivity to primary length. We shouldn't try to fix this, because it appears to work in our favor. Evidence to date suggests that single-plane cranked V-8s that have the same crank angle exhaust discharge pattern as a four-cylinder engine make less horsepower and are a little more sensitive to primary pipe length. Dyno tests with headers having primary lengths adjustable in 3-inch increments show that lengths between 24 and 42 inches have only a minor effect on the power curve, although the longer pipes did favor the low end.

Since a typical street header can have pipes ranging in length from 24 to 42 inches, we safely can conclude that each

pipe comes into its own somewhere in the used RPM range, thus helping to spread the powerband.

At this point we can sum up the situation concerning header pipe length by saying that, under most circumstances, it's not critical. This is fortunate because it's difficult to change header primary pipe length, and custom headers are expensive compared to off-the-shelf items.

With that in mind, let's move on to the secondary or collector length. Few racers pay any heed to the collector length, which is a little ironic as collector length and diameter can have much more effect on the power curve than the primary length and, on top of that, it's easy to adjust. A basic rule on collectors is that short, large diameters favor top end, while long small diameters favor low end.

Except for the most highly developed engines, most headers seen at the track are both too large in diameter and too short. For a 7,500-rpm race-cammed small-block Chevy, a collector length of 8

to 12 inches proves to be about the most effective. If the vehicle concerned has a relatively tight converter and launches at low engine RPM, then a longer secondary pipe can cut ETs by getting the car underway quicker. On a tight converter car (2000 stall) I have successfully employed a 40-inch collector length.

Headers and Quality

When I first started modifying street (as opposed to high-dollar race) small-block Chevys in the mid-1970s, an area of disappointment was the lack of affordable, well-made headers. Maybe affordable and

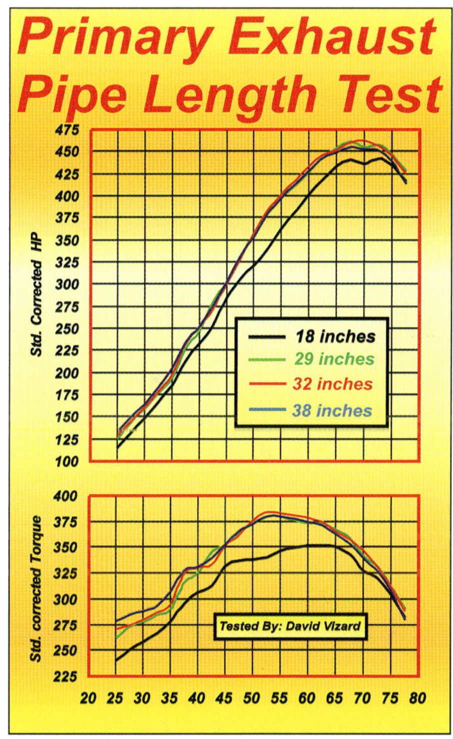

From these test results it can be seen that headers with around 18 inches do not perform as well as those with longer primaries. However, once the headers primary length is greater than about 24 inches, the system becomes very insensitive to length changes. This means we do not have to worry about getting all the primary lengths exactly equal because almost any length between 24 and 42 inches gets the job done nicely.

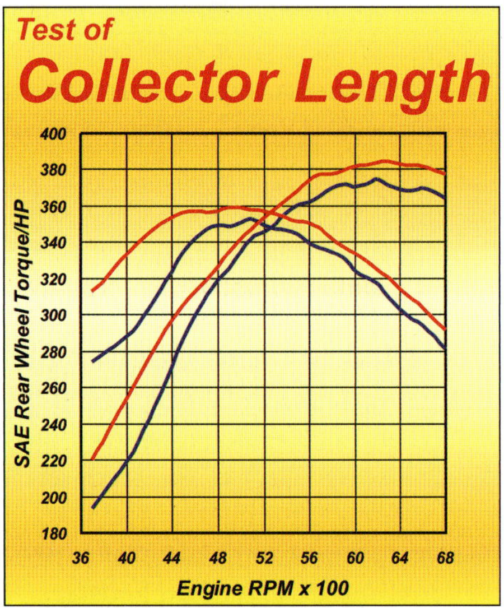

Unlike the primary length, the collector or secondary length is far more critical. What you see here is the difference in rear wheel output on an IMCA dirt car engine. The dark blue curves are for a stub about 1-inch long on the collector. The red curves are for a 10-inch secondary pipe on one side (that's all there was room for) and a more optimal 14-inch on the other. Adding these secondary lengths cost less than 15 bucks and represents a terrific return in terms of output gain per dollars spent.

What you see here is the 10-inch length added to the one side of the INMCA dirt car used for the nearby length versus output tests.

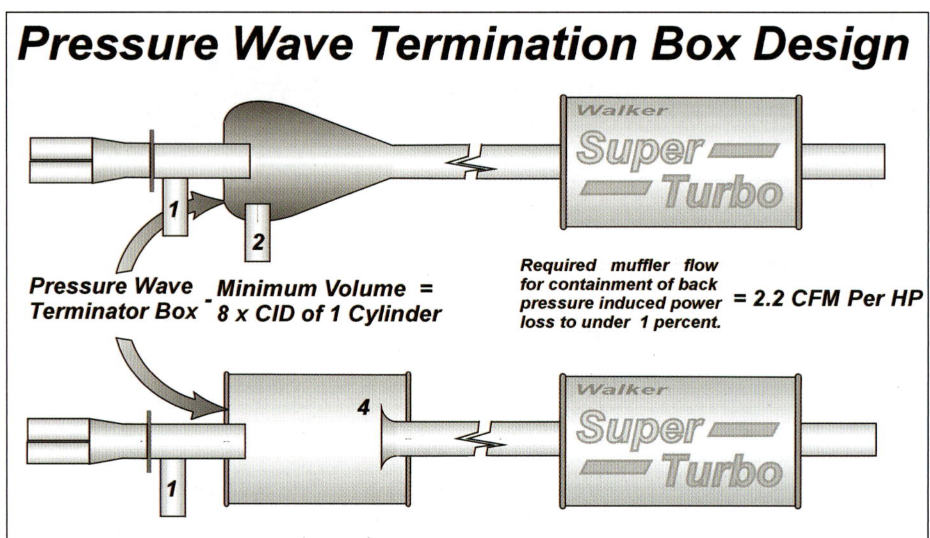

To get a zero loss system, it is of paramount importance to deal with pressure wave tuning and flow requirements as separate issues. A system designed along the lines shown here does that and can return a zero power loss while silencing down to more-than-acceptable street levels. Using these techniques, we could actually muffle a Pro Stock car down to legal street requirements without losing more than 1/2 percent of its total output! The termination box allows the secondary to be tuned to the correct length without interference from the muffler system. With the flow capability of 2.2-cfm per hp the backpressure is reduced to a level that does not have a significant effect on output.

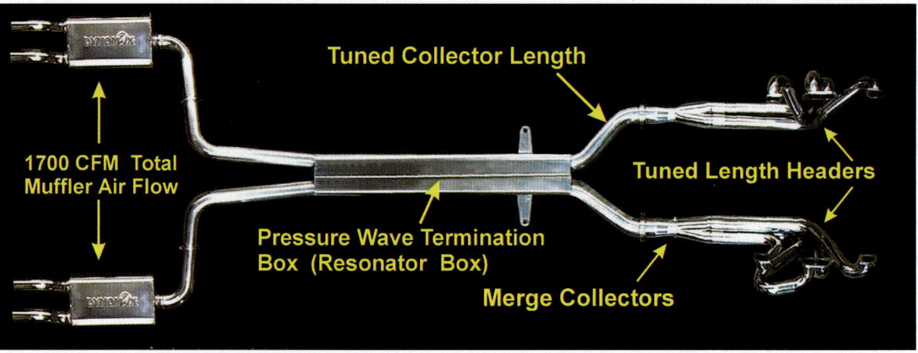

Here is a system I designed for a big-inch small-block. On what was essentially a 700-plus-hp engine the power loss was zero! Noise levels, though on the crisp side, were more than acceptable to enthusiasts and more to the point, neighbors.

well made are incompatible, but the majority of cheap headers were, in those days at least, nothing less than shoddy. My normal routine was to acquire a set of headers appropriate to the vehicle, then take them back to the shop and blast off the paint around all the welds. Next I'd inspect the welds and re-weld as required. This normally meant fixing several welds on the flanges and welding in the center of the collectors. Although the quality of less-expensive headers has improved considerably, it's still possible to buy junk headers. If you're looking for the cheapest deal, be prepared to put in several hours of work on detail fixes. These days I tend to deal with Hooker for any applications it has headers for. Modern equipment and quantity production allows Hooker to produce a good product at a reasonable price. Although far from the most expensive, Hooker isn't the cheapest; but in terms of value for money, they are hard to beat. From the forgoing you shouldn't infer that Hooker is the only company producing quality headers. There are others, but you'll have to do your own research. I have largely stuck with Hooker once I found they made functional, cost-effective parts.

Muffled Systems

Tuning the collector/secondary pipe length is all well and good, but what if the vehicle has to run a muffled system? The good news is that it's not, as is so often thought to be the case, necessary to give up or even significantly compromise exhaust system performance. To understand how this can be done, we need to consider the two major aspects of exhaust system performance separately.

These two aspects are: 1) pressure wave tuning from length selection; and, 2) minimizing backpressure by selecting suitably high-flowing exhaust components such as mufflers.

Usually, a quiet exhaust system and power are considered mutually exclusive, but this needn't be the case. It's just a question of knowing how to select suitable components and putting them all together in an appropriate manner. An important aspect toward putting together a performance system with effective muffling is muffler flow, so we'll now deal with this in detail.

Muffler Flow

When buying a carb, do you ever ask how big the venturis or butterflies are? More than likely not. After all, the engine isn't affected by the physical size of the carb, only its flow capability. If this so obviously applies to the intake side of an engine, ask yourself why the exhaust side should be any different. The truth is, it shouldn't.

Buying a muffler based on pipe diameter has no merit other than it allows the identification of a convenient

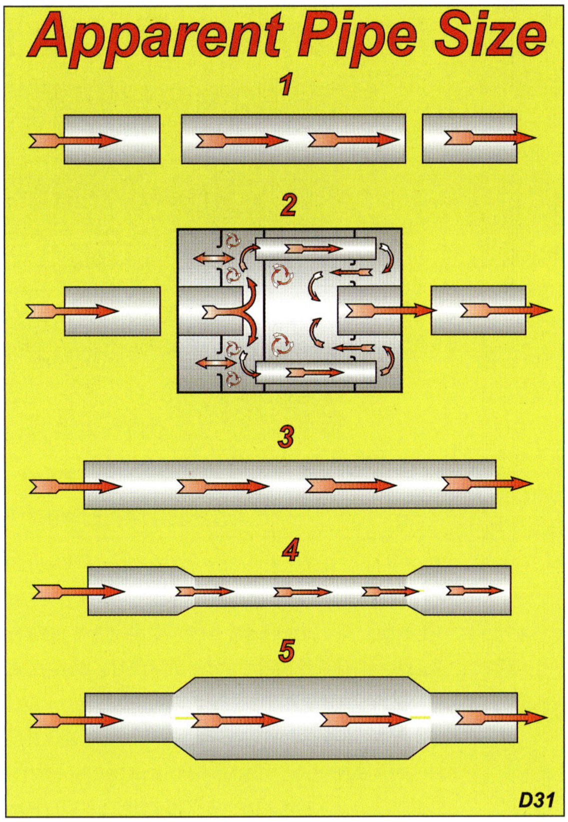

Fig 9-3. Don't confuse what it is you are trying to achieve when selecting a muffler. The size of the inlet/outlet pipe has little to do with what the engine "sees." Take a look at No. 1 set of pipes. The pipe ends represent the inlet and outlet. If whatever muffler used was 100-percent flow efficient for its size rating, it would appear to the engine as a straight pipe of the same size as the inlet and outlet pipe as per No. 3. In point of fact most mufflers are as shown in No. 2 and to the engine they appear as per No. 4. As you can see, the inlet and outlet are hardly restrictions here. What we need is a muffler with a core flow greater than the inlet and outlet pipe. If this is achieved we get something that can be represented by No. 5—and that's the best for power.

pipe size. It has little bearing on satisfying the engine's flow requirements. In case you find that hard to believe, let me tell you that there are, or have been, several well-known brands of mufflers where the only difference between a 2-inch and a 2½-inch muffler is the size of the incoming and outgoing pipe. The flow, which is largely dictated by the design of the innards, was only marginally better with the larger pipe size. Indeed, there are muffler manufacturers who have "tuned up" their products by introducing new models with larger inlet and exit diameters. The innards stayed about the same and so did the flow numbers. In such an event what was achieved? Just more muffler sales to street rodders and racers who understandably assumed bigger must be better. Just so we know exactly where things stand, let's make it clear that the engine doesn't have a tape measure and is, for all practical purposes,

totally insensitive to size. On the other hand it's sensitive to flow capability, so we should buy our mufflers based on flow, not pipe size.

In the instance just mentioned, the muffler had an increase in the entry and exit pipe, but the apparent size of the muffler core remained unchanged. It makes for an interesting mental experiment to view a muffler installation as three distinct parts. These are: the entry pipe, the muffler core, and the exit pipe. If the muffler core flows significantly less than an equivalent length of pipe the size of the entry and exit pipe, then the engine "sees" the muffler as if it were a smaller and consequently more restrictive pipe (Fig 9-3).

If the core has more flow than the equivalent pipe size, it appears larger than the entry and exit pipe. Result: the muffler is seen by the engine as a near-zero restriction. A section of straight pipe the length of a typical muffler flows about 115 cfm per square inch. This means a 2½-inch pipe will flow about 560 cfm. If a muffler of some 400 cfm is attached to a 2½-inch pipe, the engine sees the muffler as a pipe having an apparent diameter of only 2.1 inches. I bring up this point because so many hot-rodders worry about having a suitably large pipe into and out of the muffler. This concern is totally misplaced, as in almost all cases the muffler is the point of restriction, not the pipe. Few mufflers have a core flow capability greater than the pipes to and from the muffler. The only examples that come to mind are some race designs produced by Borla. The fact that the core flow is normally lower than the connecting pipe can be offset by installing a higher-flowing 3-inch muffler into a 2½-inch system.

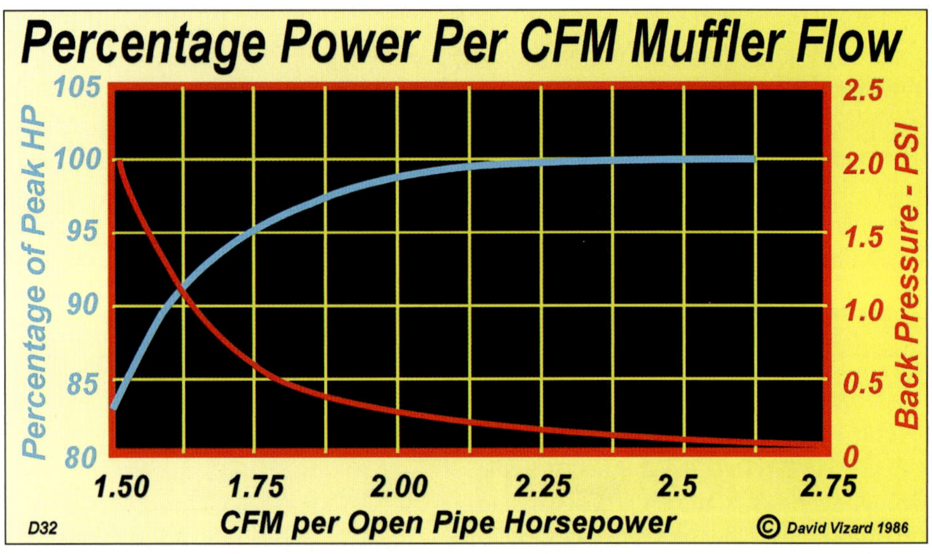

Fig 9-4. Here are the results of the backpressure tests I ran to see just how much flow is needed by an engine so it does not suffer backpressure-induced losses. As you can see, 2.2-cfm per horsepower muffler flow allows the engine to produce 99.5 percent of its open exhaust output.

Setting Functional Boundaries

Although muffler flow is an important factor, its uninhibited pursuit must be tempered by the fact that most collectors/secondary pipes are sized to produce the desired pressure wave characteristics. For instance, a 600-hp small-block may have a 3¾-inch-diameter collector that was optimally sized on the dyno. This diameter in conjunction with the length used resulted in the system tuning in at the desired RPM. But from the standpoint of flow, a 3¼-inch pipe from each bank would be capable of handling all of such an engine's flow requirements. This is yet another argument for not using pipe size to determine requirements when adequate system flow is the ultimate goal.

For many years I lobbied numerous muffler manufacturers, in U.S. and Europe, to rate their mufflers in terms of CFM so the user could make a more informed choice. A number of the manufacturers I dealt with were competent at producing good mufflers, and they sold all they could make. As a result, they saw the adoption of a new rating standard as an inconvenience. Others whose mufflers were less than satisfactory saw the adoption of a flow rating system as an extremely bad business move, because it would force them to publish figures that did little credit to their design and manufacturing capability.

Things stayed much that way until Walker got into the high-performance business. As one of the world's largest original-equipment suppliers, they had the R&D capability to design and produce functional high-performance mufflers. However, their performance-marketing personnel were frustrated by the fact that racers bought on the basis of size rather than flow and efficiency, and were often paying big money for less-effective mufflers. One could hardly place all blame on the racers for doing so. Up until Walker's decision to simplify things, the means and standards by which muffler performance had been quoted varied greatly between manufacturers and was just short of total confusion. This proved a key factor toward Walker's adoption of a CFM muffler rating. It was presumed that as an industry leader, they could influence the rest of the industry to follow suit, especially if the standards received SAE recognition and approval. This would make life easier for the user and, in a subtle manner, expose the sub-standard products of the less-competent manufacturers. The consumer benefits on both counts.

After deciding that flow, not size, was the way to rate mufflers, Walker then had to confront the issue of the test pressure used. Vast amounts of flow testing are done in the cylinder head industry, but the test pressures used range from 10 inches of water to 10 inches of mercury. In the carburetor industry, things are more organized: two-barrel non-performance carbs are rated on 3 inches of mercury (approx. 1.5 psi.) and 4-barrel carbs on 1.5 inches of mercury (approx. 0.75 psi.). Since the mufflers in question are high performance, Walker opted to adopt the same test pressure as a 4-barrel carb so all its mufflers are rated at 1.5 inches of mercury.

Required Flow

Knowing how much a muffler flows is a good start. In the absence of knowing any better, we could assume that the greater the flow the better. As it happens, in much the same way as air filters, this is one of the areas where too big doesn't hurt power. Having said that, we must not lose sight of what's going on here. Increasing muffler flow unlocks potential engine power. Once all the potential power is unlocked, further increases in exhaust system flow will not produce any further benefits in terms of power. On the other hand, depending on the muffler design, any excess flow capability may lead to an otherwise noisier system. From this we can conclude that too much muffler flow serves no useful purpose and well could cost more money than necessary. The trick is to use just the right amount of muffler; no more and certainly no less. This allows the full power potential of the engine to be realized at the lowest cost without undue compromises in terms of noise. Now the question is, how much flow is enough?

Some years ago, in anticipation of the fact that eventually almost all race cars would need to be equipped with mufflers, I set out to determine how much flow was enough. Initially such tests look easy to perform, but the degree to which the pressure in the combustion chamber is reduced or scavenged during the overlap period is a function of both pressure wave tuning and system backpressure due to available exhaust system flow.

To determine the effect of muffler flow alone on power, it was necessary—as far as possible—to isolate the effects of flow from the effects of pressure wave tuning from pipe sizing. This can be done with a pressure-wave terminator box system shown earlier that, as it happens, is an intrinsically important part of building a high-performance system. Let's look at some test results.

In Fig 9-4 you see the results of tests run on a number of engines of various types. The only fact of significance they had in common was the use of a cam with 290 degrees or more of seat duration. The trend is that, as flow is added to an initially flow-restricted engine, power increases rapidly at first then the gains

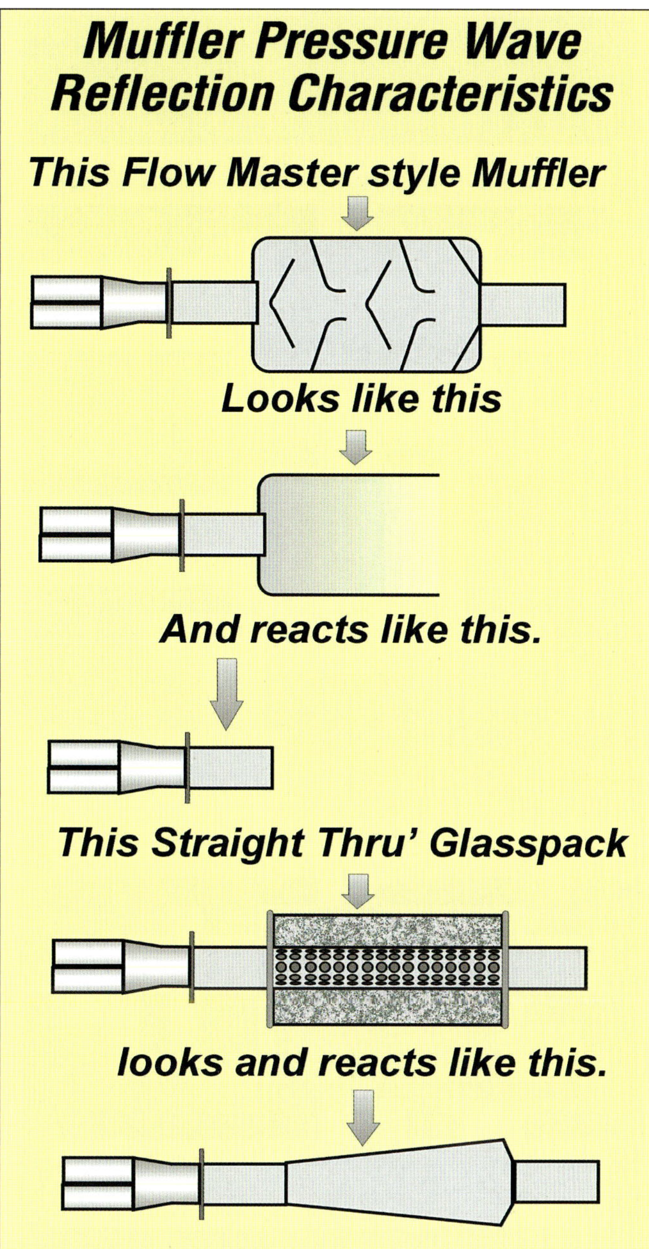

Fig 9-5. A muffler test where the mufflers capability is tested just by attaching it to the end of the collector is a false test. Here's why: We already know that the engine's output is sensitive to the secondary lengths. Just adding a muffler without consideration of it's type can drastically alter the secondary tuned lengths and in so doing completely invalidate the test.

seen tail off. Once the flow exceeds about 2.2 cfm per horsepower, the gains from increased muffler capacity drop to about 1 percent or less. In other words, the engine's demand for flow has been satisfied and backpressure-induced power losses are contained to within less than 1 percent of open-pipe output. With the availability of this key number, you're now in position to determine how much muffler flow your engine is likely to need. It only requires that you make a reasonable estimate of its open exhaust power

potential, and then multiply this number by 2.2. For example, a small-block Chevy that made 400 hp on open exhaust will require 880 cfm (400 x 2.2). Two 440-cfm mufflers will get the job done and contain the loss to 4 hp or less. See how easy making an appropriate choice becomes with mufflers rated in CFM?

Pressure Waves

Having learned how much muffler flow is needed per horsepower, we can move on to methods of applying suitable-capacity mufflers to the system without needless disruption of length-induced pressure wave tuning. Probably the best way to ease into this somewhat complex subject is to consider muffler test results done in recent years and published in various monthly publications. These tests have shown, apparently to the total satisfaction of those conducting them, that sometimes low-flow mufflers developing at least some backpressure were required to make the best power. Unfortunately, the conclusions (as opposed to the tests) I've seen in all such tests were not valid. There are several reasons contributing to this, and they're relevant to our quest for a low power-loss

Here UNCC motorsports students Nate and Justin check the flow of a big Borla muffler. Their findings— big flow numbers!

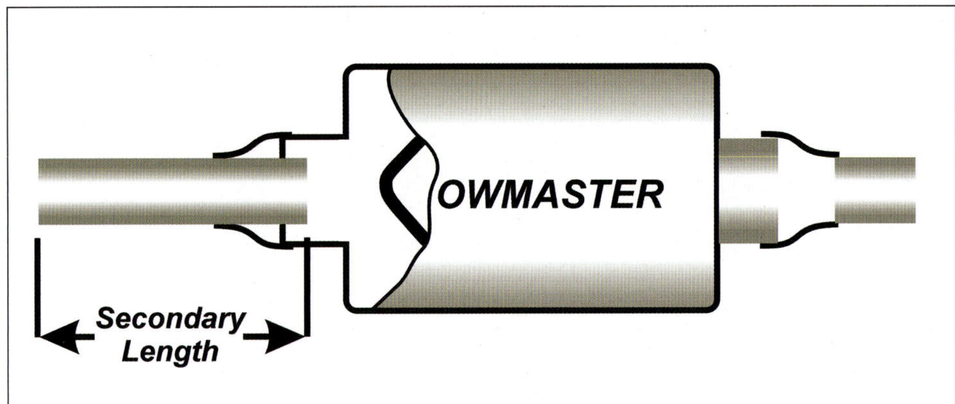

Excellent results have been seen with a large-bore high-flow Flowmaster when used as in this diagram. Not only does this layout allow for good pressure wave tuning, but with the entry style used there is a measure of anti-reversion that reduces the positive pressure returning up the exhaust. The result is a small but measurable increase in low-speed torque.

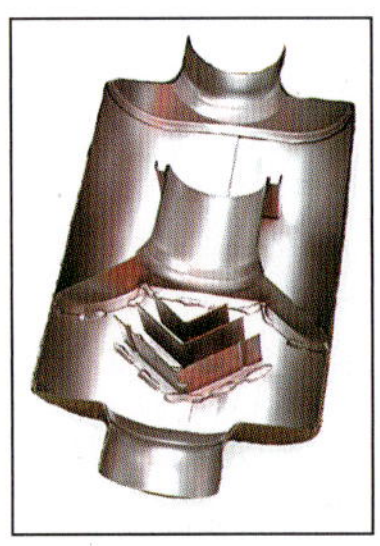

The Flowmaster muffler is a classic example of an open muffler in as much as the exhaust thinks it has, to a large extent, arrived at open atmosphere.

The Walker Dynomax is an excellent low cost muffler that should be given serious consideration if you are working on a tight budget. The three-pass design is the result of some good flow bench work and as a result flow is far better than most mufflers of this type. Noise reduction is also a little better than average for a high-performance muffler.

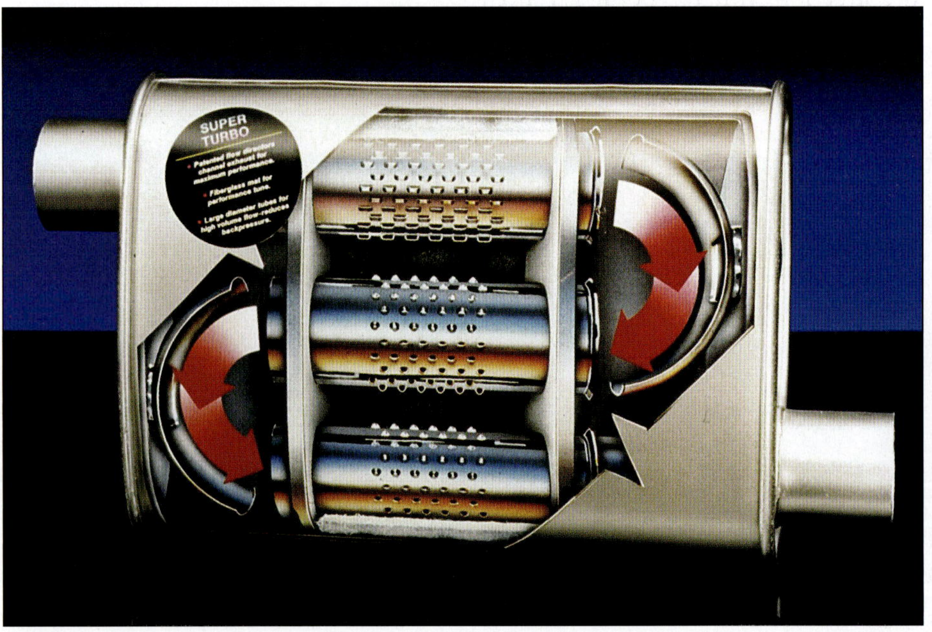

exhaust system. The first is that the internal design of mufflers varies greatly.

Many mufflers are made up of a number of inter-connected chambers that give the exhaust varying degrees of access ease. Others are of the "glass pack" variety. These types represent opposite ends of a spectrum and have a substantially differing response to arriving pressure waves (Fig 9-5).

Earlier on it was emphasized that the collector length was, in most cases, more influential than the primary pipes. Adding a muffler to a system with already optimized lengths may produce a pressure wave-induced response that has a far greater effect on power than the change in muffler flow.

Let's assume that the test muffler is attached directly to the end of the collector. Remember, a pressure wave is reflected when it reaches the end of the exhaust pipe or when a substantial increase in cross-sectional area occurs. Mufflers with open chambers, such as Flowmaster's, often appear to the pressure wave much the same as the end of the pipe. This means the pressure waves see no change in length and reflection occurs largely as it did prior to the fitment of the muffler. Now let's look at the situation with a hypothetical high-flow glass pack. If the glass pack is densely packed and the perforation holes are small, the muffler can appear as a considerable extension of the tailpipe length. A 3-inch change in tailpipe length can have a measurable effect on the power curve, so what would you expect if another apparent 18 inches was added? Answer: A big drop in top end power, and if tests started at 5,000 rpm, the potential gain in low end would never be seen because it would happen below the test's starting RPM. This and other interrelated factors imply that many comparative muffler tests were in fact pseudo pipe-length tests. Although many conclusions drawn from such

This straight-through stainless Walker Dynomax muffler may not be the prettiest you will see, but it is very functional. It resists corrosion, produces an authoritive sound, and allows power to be made with minimal loss.

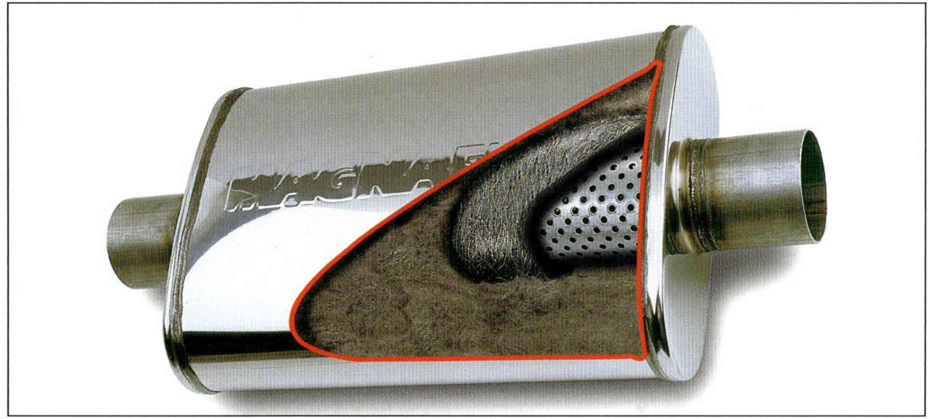

Magnaflow is another muffler design that produces good results when correctly sized CFM-wise to the engine concerned. The exhaust note produced is also pleasing to the hot rodders ear!

tests proved invalid, the tests still demonstrated some important truths. The principle truth is that flow requirements and pressure-wave-length tuning need to be isolated from one another as far as possible.

This can, to a large extent, be done with the pressure wave terminator box mentioned earlier. The terminator box makes everything downstream appear near invisible to the header's primary and secondary tuned lengths. With a flow capability of 2.2 cfm or more, the muffler appears virtually invisible from the standpoint of flow induced backpressure. As a result, we have a muffled system that produces virtually the same power as an open exhaust.

Noise Abatement

So far we've talked about limiting the power loss usually suffered when a muffled system is installed onto a high-performance engine. Through all this we mustn't lose sight of the fact that the point of a muffler is to muffle.

By using no more muffler than needed, you're at least giving the mufflers selected the best chance of doing the job. Unfortunately, mufflers can be a little inconsistent and unpredictable in terms of noise suppression from one engine type to another. But we're dealing only with the small-block Chevy engine, and much testing has been done for this application. Even so, it's possible that a system that

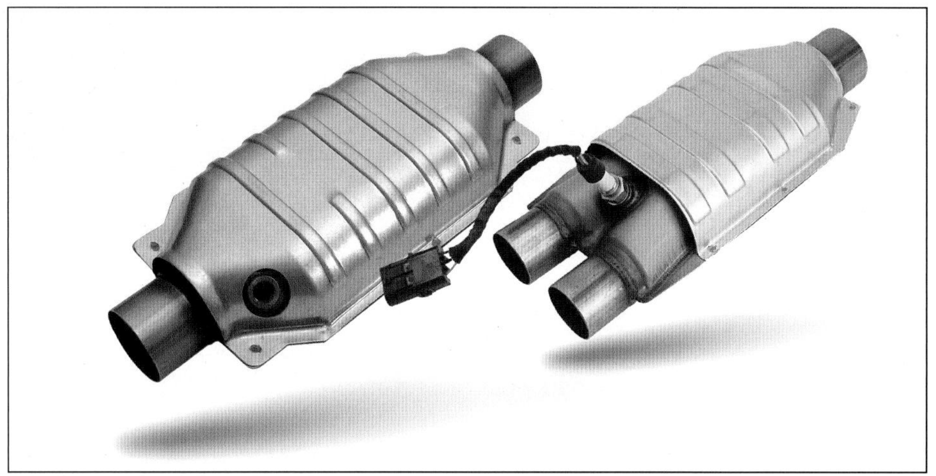

Be aware that selecting a high-flow muffler will be of little use if the cats used are low flow. Check out what's available here from Walker and other high-flow cat manufacurers.

produced acceptable results on one engine is less so when the engine combination changes. Situations involving high compression ratios, long period cams, and nitrous oxide are usually more demanding in terms of noise suppression. Big cubes, shorter cams, and lower compression ratios are easier to muffle. Unfortunately, aftermarket performance muffler noise levels are rarely comparable to original equipment. Still a little more noise from a nicely tuned engine is, for a performance enthusiast, just another form of music.

The biggest problem in this area is finding you have put together a combination that's not quiet enough. Since a muffler's noise suppression capability varies with the engine's basic sound signature, it's difficult to precisely rate the mufflers' sound suppression capability. In this area, Walker once again is attempting to set the pattern for the future of the performance muffler industry. They are doing this in two ways: first, the mufflers have the flow rating right on the box; and, second, they have sound recordings of various mufflers on a performance engine on their website at Walkerexhaust.com.

Tailpipe lengths can also influence the results achieved with a muffled system. As far as power is concerned, tail pipe length after the muffler has no measurable effect on output if a large change in cross-section is present up-steam of the tail pipe. An open type muffler or a terminator type box would constitute such a cross-sectional change. The tail pipe length from most glass packs is also of little consequence if a cross-sectional change is present up-stream. The tail pipe length can be significant from the noise suppression aspect. Don't let the tailpipe end under the car. Either have them go all the way to the rear with down turned exit pipes angled slightly in toward each other, or have side exits aimed 45 degrees toward the ground. The rear exit though, is the preferred way to do the job.

Balance Pipes

The definition of a balance pipe is a pipe connecting the collectors of one bank of cylinders with the bank opposite. Balance pipes have two possible attributes: increased power, and reduced noise. My experience dyno testing both these factors has shown that a balance pipe has proven 100-percent successful at reducing noise. The reductions amount to a minimum of 1 dB, with 2 being common.

This is a worthwhile sound reduction especially as it could make the difference between passing and not passing tech. As for power increase, these prove less certain than the noise reduction. In about 60 percent of the cases I've tried, the engine responded by delivering up to about 12-hp more with 5 to 8 being about the norm. In the remaining 40 percent the power was virtually unchanged. Although I have tested balance pipes on about 40 different small-blocks ranging from hot street to race, I have never seen a reduction of power.

This means the use of a balance pipe is worthwhile because it always reduces noise and more often than not increases power. That means there is no downside.

The dimensions of the balance pipes are not overly critical. The only dimension that appears to have measurable influence is the pipe diameter. This requires an area at least equal to that of a 2¼-inch diameter (4 square inches), with 2½ to 2¾ being preferable. Anything above 2¾ inches does not appear to deliver any further gains, but I have only conducted tests on engines up to about 600 hp. As for the length of the balance pipe, this appears to be largely immaterial. Dyno tests indicate balance pipe lengths as short as 18 inches respond in virtually the same manner as ones 72-inches long.

Summary

Although I have not dealt with the theory behind everything covered in this chapter, you still should be sufficiently well informed to assemble a near-zero-loss muffled exhaust system. As long as you don't lose sight of the essentials and principles involved, good results will be achieved. These guidelines are sufficiently well defined to accomplish the desired goals. Step outside these recommendations and you are on your own!

IGNITION SYSTEMS

For our budget-oriented purposes we find that conceptually at least, GM's HEI distributor is a near ideal choice in terms of practicality. However in terms of high-RPM function, it falls well short of ideal. But before looking at the possible fixes lets consider why we might find it best to stick with an HEI-style distributor instead of just swapping out the entire ignition system.

First just ask yourself where else you can find a distributor and ignition system packaged into one compact, easy-to-use assembly. Installation involves little more than just dropping the unit into the motor and plugging in one wire. It's hard to imagine anything much simpler, and on that basis, the HEI deserves high marks.

Because the HEI is such a simple ignition system to use and has the potential to service a 9,000-rpm race engine, I'm not going to delve into endless other possibilities here. If you want to stray from my recommendations here, feel free to do so but consider this: if you are going to do your own R&D, it's just that and you are on your own!

Before going into a lot of details about what you can do and what you should do let me set the playing field here. There are a number of companies who are in the HEI business and I pretty much know all the bosses of all these companies. I say this so

There are a lot of fancy ignition systems available for the small-block Chevy but the simplest and most cost-effective is still the HEI—but it needs a little hopping up to get the job done. A custom built Performance Distributors unit takes care of that.

you understand the fact that I am in a strong position to test all of the units involved. These companies are MSD, Accel, Pertronix, and lastly, Performance Distributors. All of these companies make good stuff, but Performance Distributors has, by far, the best and most personal customer service. Performance Distributors can and will deal with every customer on a one-on-one basis and that makes it very convenient for you to precisely get what you and your engine need. You will pay a little more when dealing with Performance Distributors but when you come up against a prob-

lem—especially ones related to ignition curves—you will be glad you did. With that said, you have the option to deal with any of the other companies mentioned for whatever parts you are looking for, but from here on out, I am going to be referring to Performance Distributors almost exclusively. Just an aside here: I have been dealing with Performance Distributors for much of my HEI ignition needs and have used its units in well over 60 magazine project engines. Never, in more than 30 years, have I had one single problem with its products. Every one has functioned just like it should!

This Scat cranked AFR headed 383 made 503 ft-lbs and missed 600 hp by just a couple of hp—all on a 10.5:1 CR and pump fuel. The sparks for all this power came from a Performance Distributors HEI.

HEI Shortcomings

Since we are dealing with budget builds, there are two factors that must be taken into account. First, your intent might be to install a used unit and it may or may not be in need of a rebuild. Second, an HEI in stock trim falls short of what we need for performance. Although outstanding in concept, the high-RPM spark capability it delivers most certainly is not. Depending on the coil and module characteristics, a stock HEI's spark energy starts to fall off at about 3,300 rpm. Although still capable of jumping a typical plug gap, the spark intensity begins to drop below what's needed for a higher-compression performance motor. At about 5,250 to 5,750 rpm, a stock HEI will drop sparks. Although a Band-Aid fix is to close the plug gaps, this isn't what's needed to make good power.

Basic Rebuild

As a first step, let's look at ways and means of sourcing an HEI with which to work. If you bought a complete motor or already have an HEI because your project engine is already so equipped, you're in a good position. HEIs don't wear their bear-ings as much as points distributors because the loading at the top is uniform, and at the bottom is well lubricated. This means there's a fair chance you can hop up your existing unit to get the job done.

If you don't already have a distributor, you have two options. First, pick up a used unit at the salvage yard and rebuild it. This will take a couple of hours and will require, as a minimum, the replacement of the rotor and cap. You also need to determine if the bearings in the body are up to further service. If there is more than just a hint of side-to-side movement it's a fair chance that the bearings need replacement. The best bet here is to start with another used unit because at least 50 percent of any pile of used HEI still have acceptable bearing clearance. Also check the advance weights. These wear at the pivot point and as a result produce an inconsistent advance curve. Check the pivot points here. If the actual posts are worn then you might consider dumping the whole unit because it could need a new shaft and that is a little pricey. However there is a point you might want to consider before tossing whatever unit you have. Performance Distributors can furnish you with a shaft with weights and springs with a custom curve already done to compliment the engines specs you have. If the bearings in the body are decent you may want to consider that route. If only the weights are worn then you can get a set of replacement ones along with springs to build the advance curve needed from Performance Distributors.

Although it will cost a little more, your second option is to buy an Accel Blue Print rebuilt unit. These go for about $150 through any of the big parts houses. At first, this may seem just a shade more costly than doing the job yourself. The Accel unit, though, is more distributor than just a stock rebuild. My own spin tests show it to be capable of at about 750-rpm more than a typical stock unit because of a superior module. In addition, it comes with an adjustable vacuum advance and springs for the mechanical advance that will allow a dozen different mechanical advance curves. In all, this unit is a good deal.

More Spark, More RPM

If the application requires more than 5,500 rpm, you'll need to replace performance-related parts such as the coil or module. That is what we will deal with now.

At this point I'll cover ways to make a basic HEI unit run to higher RPM. If you're expecting dyno test results in this chapter, forget it. It's already a proven fact that bigger, fatter, hotter sparks are unlikely to hurt power. For this reason I'll rely on bench tests to show what can be done to improve the stock HEI to full race capability.

With an HEI distributor, the critical components for a big, high-RPM spark capability are the coil and the module. To perform the tests shown in this chapter, the HEI's spark output was dumped into a chamber pressurized with pure nitrogen at some 142 psi. Although this can be considered a valid A-to-B comparative test procedure, there are some points to note so you can more precisely relate results to actual use in an engine. Pure nitrogen is a far better insulator than air, which basically consists of 20-percent oxygen and almost 80-percent nitrogen. Also, cylinder heat (which was not simulated in our tests) makes it easier for the spark to jump the gap. This means it will be harder for an ignition system to fire the plugs in the test rig than in an engine. As a result, an ignition system starts dropping sparks in the test rig at about 15- to 20-percent sooner than it does in a running engine. The results you see here have been corrected for this.

Coil Test

Although the coil's turn ratio for high-output voltage is a prime concern, there are two other factors to consider such as the total current draw and the rate at which charging and discharging take place. Without a quick enough charging time, a coil won't fully charge between sparks at high RPM. A more rapid charging rate can be brought about by reducing coil inductance, which usually means fewer turns on the primary side. Making such a change proves to be a double-edged sword. Although reducing the primary turns increases the primary-to-secondary turns ratio, thus producing a higher secondary voltage, it also increases the current draw. The coil design must be the result of a workable compromise. After testing, it was found that the rate of current built up in the stock coil was 2.5 amps per millisecond. The same test on a good aftermarket coil showed it to be faster at 3 amps per millisecond.

To see how this and any other coil design differences affected the spark capability, an otherwise-stock HEI was spin-tested first with the stock and then the MSD, Accel, and Performance Distributors coils. The stock HEI ran to a spark-dropping limit of 4,800 rpm. Replacing the stock coil with the MSD coil raised this to 6,100 rpm. The Accel and Performance Distributors coils, within 1 percent, duplicated the MSD coils performance. This test then shows that all these coils are capable of delivering a spark-limited RPM increase of 25 percent or more over stock (Fig 10-1).

Swapping out the stock coil for a high output aftermarket coil is an easy 10-minute job. It's even easier if it is done at the same time a new cap and rotor is fitted.

Module Test

The function of the module in an HEI is much the same as contact breaker points in an older-style distributor. Its first job is to supply some 200 to 300 volts to the coil primary and switch it on and off at the appropriate time. The best module is one that reliably handles the highest current possible, delivers the highest voltage to the coil, switches as fast as possible, and has the longest dwell time (switched-on time) possible.

I tested all these factors on a high-performance module, but the most significant differences between this and a stock module showed up in the current and voltage delivered to the coil. A stock GM module typically delivers a maximum of some 5.5 amps. By comparison, the high-performance module delivered 7.5 amps, so its prospects for delivering a strong spark at significantly higher RPM obviously looked good.

The next tests we're going to look at relate to the module's ability to deliver

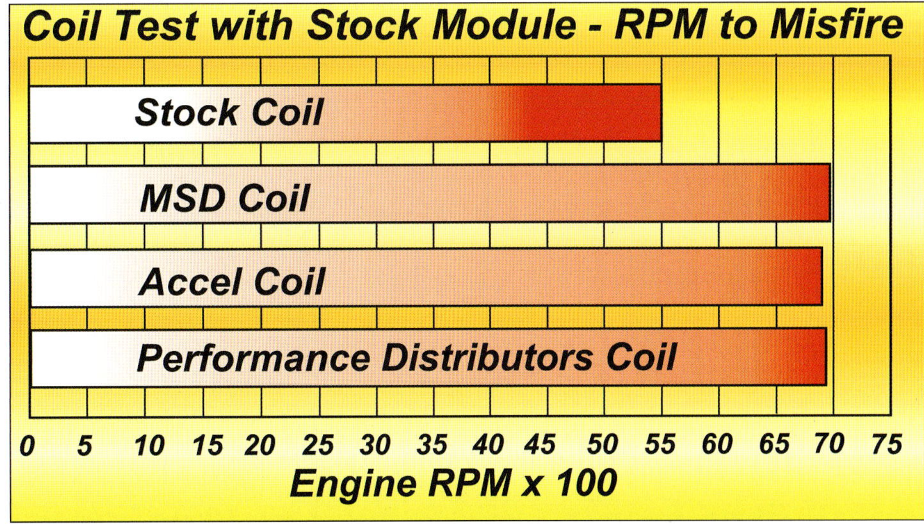

Fig 10-1. This test shows what a typical aftermarket coil can do to increase the spark limit RPM of an otherwise stock HEI (corrected from test rig to engine RPM).

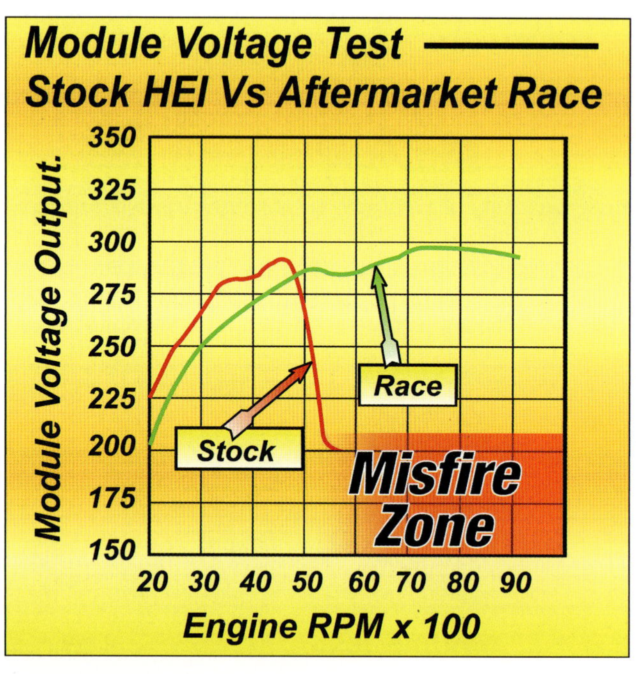

For really high RPM, it is necessary to upgrade the module. Shown here is the Performance Distributors Dyna-Mod unit.

voltage to the coil. The winner in this event will be the module that produces the highest voltage up to the highest RPM. For the results, check out Fig 10-2. As you can see, the high-performance aftermarket module wins by a healthy margin once again. The test results shown in Fig 10-2 show what's required to produce more spark at higher RPM. Modules from Performance distributors, Accel and MSD all seek to do this.

Spark Test

The spark-producing components for an HEI from the companies I am mentioning here certainly prove their worth individually, but the real question is what they're worth collectively. A 4,800-rpm limit on the test rig was equal to 5,500 on a real engine, so I will use this ratio for some real-world RPM capabilities. But there is a factor I need to throw in here. The stock HEI numbers were with a stock-size plug gap of 0.040. A high-performance engine responds to an increase in plug gap because the wider gap produces a more energetic and powerful spark. The tests for the upgraded HEI were with 0.055 plug gaps. The Performance Distributors coil and module equipped unit was good for a solid 9,200 rpm (Performance Distributors claim

9,000). The Accel components drove the HEI to a little over 8,900, and the MSD components drove our test HEI to the limit of our spin machine, which was 9,600 rpm.

As tempting as it may seem to use the system that gave the highest RPM, you should be aware that on the way to the rev limit there was no measurable difference in the sparks. This means if you intend to build an engine that turns only, say, 7,000 rpm, a system that goes to 7,500 will get the job done in exactly the same fashion as one that goes to 9,500. The point I am trying to make here is you don't need to buy any more RPM than about 500 above what you perceive is needed. It's OK if the system you choose goes more, but there are no benefits to be had other than you know the system is more than up to the job.

Plug Cables

Assuming you have prepped the distributor as suggested with at least a high-performance module, then your plug cable requirements for an affordable alternative start to look good. The worst situation is to use old carbon string-type leads. These tend to decay in capability and cheap ones are often the source of a sizable power loss when they've seen 40,000 miles or two years. However, such cables, when used with the improved capability of a modified HEI, will deliver results as good as the high-quality, wire-wound inductive suppressed cables from Accel, MSD, Moroso, etc. The only difference is they won't last, but they will get the job done for a year at least. For a reliable name brand at a low price, check out Accel's Super Stock cables.

Mechanical Attributes

Once a high-energy spark can be generated at sufficiently high RPM, the next issue to address is ignition timing and advance curve characteristics. Knowing what controls the advance curve and how to change it is almost common knowledge.

The mechanical or RPM-generated advance curve is controlled by the weights (white arrows) and springs (yellow arrows).

cams are used, the low-speed compression pressure is reduced, so more advance is required until the dynamic effects take over and start filling the cylinder to significantly higher levels.

Let's go through some examples and see how this works. For a starter, let's assume we have a short-cammed 9:1 truck motor. The early closing of the intake valve will mean a lot of charge is trapped at low RPM; this, in conjunction with the higher-than-stock CR, will produce a cranking pressure of around 180 or more psi. At low RPM, the charge will burn faster than in a stock unit. Result: this engine now needs less initial mechanical advance, and the total required will probably be less, so the mechanical advance will only need to come in slowly. Initial timing may range from 5-degrees BTDC to zero and total around 28 to 32 at the most degrees. When the engine is cruising at part throttle, the pressure just prior to ignition is substantially reduced due to the fact that the engine is throttled. To get fuel efficiency under such circumstances, it's necessary to advance the timing considerably. At 60 mph in high gear, a typical small-block Chevy needs between 45 and 55 degrees of advance. Because of the short cam, the vacuum advance in

The big questions are: what to change it to, and why did it need changing?

The grassroots answers are: cylinder pressure prior to ignition, and, to a lesser extent, heat. Contrary to popular belief, a charge doesn't explode, but burns progressively. The best output is usually achieved when peak pressure occurs at about 15 degrees after TDC. To achieve this, it's necessary to ignite the charge well before. The compression pressure significantly affects the speed of the burn. The higher the pressure, the faster the burn rate. This means that as the CR is increased, the ignition advance required is reduced. When longer period

Trimming the advance curve is usually just a question of changing springs. The plastic screws shown here replace the metal ones that hold the rotor in place. These avoid any chance of a spark jumping from the rotor screws and firing the cylinder at the wrong time.

Here is how the vacuum advance operates. When the rod that comes out of the "can" and up through the pickup point indicated by the yellow arrow is pulled toward the can under vacuum conditions, it rotates the base (white arrow). This advances the timing.

---

our current example will not need to come in until about 7 to maybe as much as 10 inches of mercury, and the amount required will be at most about 18-degrees distributor (36 crank). This, added to the slow mechanical advance occurring at 2,000 to 2,500 RPM, should result in about 45 degrees under these conditions.

If a relatively large cam is used, then the advance needs to come on faster at first, although little or no additional total may be required. If the cam was used with the CR's recommended, the cranking pressure won't be significantly different. This means the quicker rate of advance needed by the longer cam can, to a large extent, be offset by the increased CR. Mechanical advance won't change significantly until race-category cams are used. The vacuum advance for a longer cam, though, will need rethinking. The longer the cam, the sooner the vacuum canister needs to start pulling in advance. A 285-degree cam may need the "can" to start applying vacuum at 3 to 5 inches with full vacuum advance in at around 10 to 14 inches. At cruise, such a cam will, even if the CR is well matched, require about 50 degrees.

Now if all these interrelated factors sound a little confusing, just remember that Performance Distributors can build what your engine needs at a very reasonable price.

This is the distributor machine that every Performance Distributors unit is tested on for curve accuracy before shipping.

Timing Controls

To achieve the timing required, we have three system controls. These are: 1) the mechanical advance, 2) the vacuum advance, and, 3) the total advance stop. Ideally, all the required advance characteristics should be determined by testing on a chassis dyno. But dyno time costs money, and though I strongly recommend a dyno setting up session even for the budget racer, it's worthwhile getting as many adjustments done as closely as possible before renting dyno time. Failing that, let's look at how the advance curve can be set up by test driving. The technique I'm about to describe only works if your vehicle is equipped with a converter that stalls at or close to typical stock levels and is muffled to the extent that you can hear detonation should it occur. If the converter stalls at much above 3,500, the effect of the ignition timing below this will affect throttle response only. The mechanical needs to be done at the track, while the vacuum needs to be done on the freeway.

Mechanical Advance

At this point I'll assume you have an HEI with a selection of advance springs and an adjustable vacuum. The first step is to install the strongest springs at both locations on the distributor's advance mechanism. It's necessary to deal with manuals and automatics differently. Automatics first. Stage the car a known distance into the staging light. It matters little what this may be, but it must be close to the same for each run. You're

An adjustable vacuum can is a boon to easier setting of the vacuum advance.

This 350 eventually made over 580 hp on pump gas and was still just about street drivable. Note the use of a Performance Distributors HEI. This got the job done on this 8,000-rpm-plus engine.

interested only in the first 60 feet, and the tires must grip well enough to prevent wheelspin. Make a pass and check the 60-foot time. Next, replace one of the springs with the one having slightly less tension. Make another pass. If the car is faster, repeat the process with a lighter spring yet. Continue with progressively lighter springs until either detonation is heard or the 60-foot times stop improving. At this point, make a pass the entire length of the strip. Now replace the second spring with one having less tension. Make another pass and check the ET. If it's faster, repeat the process until the best is established.

The advance curve now could be anywhere from 90 percent as good as it can be to the best possible with a mechanical distributor. It takes springs, weights, and a dyno to get really close. What we haven't dealt with at this stage is the total advance. This affects top speed more than anything, and if the tests so far were done at 28 to 30 degrees, the total will be short of optimum. If the whole distributor now is advanced, the low-speed initial advance may be too much. Go back into the dis-

tributor and replace the light primary spring with one that's slightly stronger. Now make passes and advance the dis-

tributor progressively until the best ET/trap speed is seen. Last, go back and evaluate the secondary spring that you first started with. You're now done, and this is as good as it gets short of a dyno.

If your vehicle is a manual transmission, all of your tests can be done on a lonely stretch of road with a stopwatch. The technique involves choosing a suitably low starting RPM and timing the vehicle between two speeds. Let's assume that the motor has the potential to pull from 1,500 rpm. With the strongest springs installed, take off and have the clutch fully engaged in second or third gear at an RPM lower than the test's starting point. For the example, we're working here with 1,200 rpm. Floor the throttle and time how long it takes to reach 2,500 rpm. Progressively lighten up the primary spring until the best 1,500-to 2,500-rpm time results. Now repeat the tests using RPM from 2,500 to 4,000. Take the car to the strip and set the total. If this is more than a couple of degrees different

The distributor used in this engine was also a Pertronix unit but not of an HEI style. It required the use of a separate coil and ignition box.

Under all those wires is an MSD distributor for use with an MSD 6 box and an ignition retard for nitrous use.

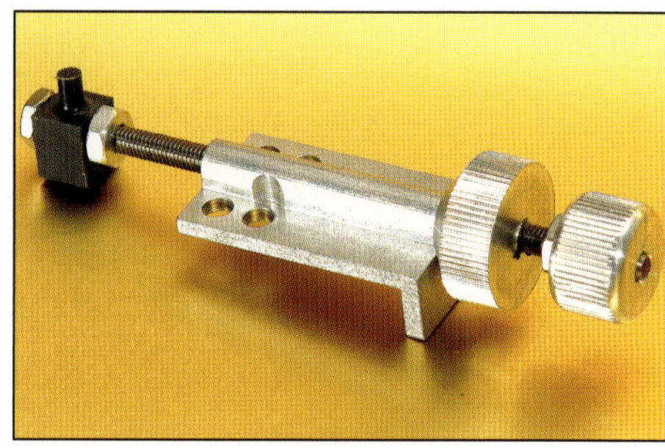

On the dyno or at the track, this micro adjuster for setting the timing is a boon. It allows precise changes in timing as small as 1/4 degree. This can be important when trying to squeeze the last ounce of power from an engine. On some occasions I have seen as much a 3 hp from a half-degree change in timing.

than the setting used while developing the advance curve and any sign of detonation is apparent, go back on the primary spring to one slightly heavier.

Vacuum Advance and Mileage

Vacuum advance is critical if good fuel economy is to be achieved. Although you may not feel this is of great importance to a race car, having vacuum advance cleans up the way the motor drives in the pits or paddock and reduces the possibility of fuel-fouled plugs. Although an adjustable vacuum looks ideal, its adjustability is in terms of the amount—not the rate—at which it comes in. It's a compromise, but it's a practical compromise. Trying a dozen cans in an effort to get the optimum wears thin after a while.

To set up an adjustable unit, start with the advance cut way down and drive the vehicle. If no detonation is heard during normal, part-throttle driving, then add a little more advance. Continue with this process until detonation is apparent and then back out some of the timing until detonation has ceased. You're now in business.

This is where the ignition has to prove itself out. If it doesn't work here, it's replaced!

NITROUS OXIDE

Nitrous oxide injection is the budget performance enthusiast/racer's surefire route to big power. No other form of power generation comes close to delivering so much horsepower for so little money. This chapter deals with results rather than going into system design detail. But first some basics.

Nitrous Oxide is a compound consisting of one-third oxygen and two-thirds nitrogen. By comparison, our atmosphere consists of one-fifth oxygen, with the rest mostly nitrogen. N_2O is a gas at room temperature, but if (in the same manner as propane gas) it's held in a container under its own vapor pressure it's a liquid. Depending on temperature, the vapor pressure is typically 700 to 900 psi. When released from the bottle, the expansion and partial vaporization the N_2O experiences cause it to super cool to –128 degrees F. N_2O is a liquid at this temperature. Using electric solenoid valves to activate the flow, N_2O largely in liquid form can be injected into the engine. Since N_2O is 50-percent richer in oxygen than air, and it's largely in liquid form, a great deal of additional oxygen can be delivered to the cylinders.

The additional oxygen supplied by the N_2O must have its own supply of fuel;

Here is a 350-inch build toward the upper end of our cost constraints. It made 768 horsepower with the 300-hp jets installed into the Zex nitrous system.

otherwise, the now oxygen-rich charge will simulate an oxy-acetylene cutting torch and melt the pistons. The additional fuel is supplied by tapping into the pressure side of the fuel line and is delivered on demand via a solenoid valve similar to that used for the N_2O.

With both solenoids connected to a common activation switch, the system can be fired by simply pushing a button.

Here's one of my low-buck engines with a set of ported EQ iron heads. With the Zex kit calibrated for 175 hp, this motor's base 490 hp jumped to 652. In case you are wondering why all my so-called cheap motors have expensive looking March belt drive systems it's because they want me to show them off. For them it's a cheap form of advertising and I get a good-looking engine out of the deal. Without all the front accessories and the headers this engine cost around $3,900 to build.

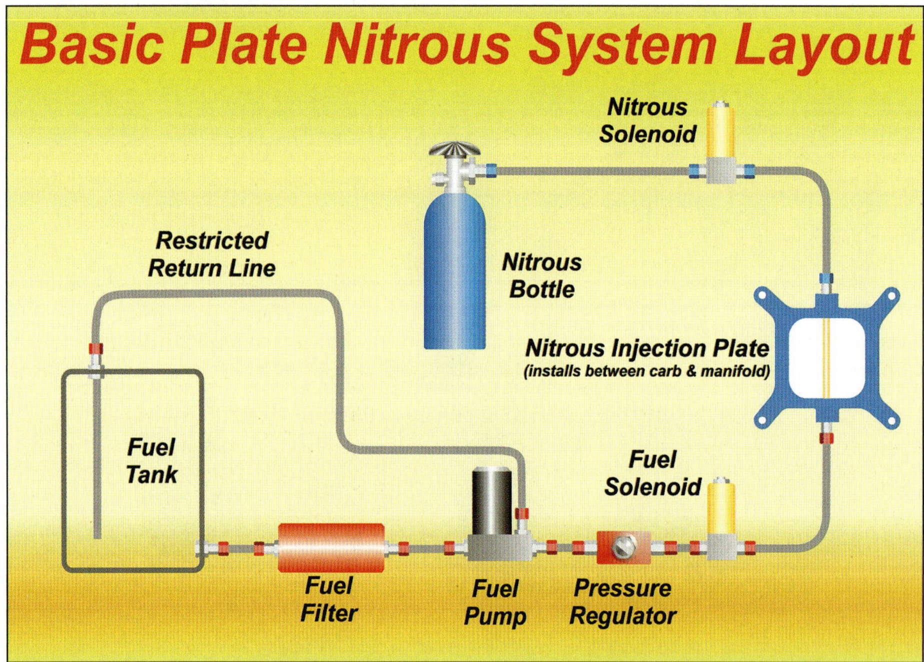

Fig 11-1. Here is a basic plate system. Although I have not shown it the solenoids are connected by wires to the operating switch. Activating this opens the solenoids and the nitrous and additional fuel is injected through the spray bars in the plate. The fuel pump shown here depicts a good high-volume aftermarket pump with a bypass back to the tank so the pump is not dead-headed. This is important if the pump is a relatively high pressure one, but a low-pressure pump at about 5 to 7 psi won't normally need this return line. Jetting for the system is at the entry into the ends of the spray bar in the plate.

Fig 11-1 shows the layout of a plate system that installs between the carburetor and intake manifold.

Effective Systems

A nitrous system can take the form of a simple plate installed between the carb and manifold or injector nozzles installed into the manifold port runners. For high-output installations, both methods are used in the form of a two-stage system. The first system to be activated may bring in 200 hp, and the second system is activated as RPM climbs. This is done to avoid extremely high cylinder pressures that would result from injecting large quantities of nitrous into an engine at low RPM.

Because we're dealing with budget constraints, only plate systems will be considered. They're less expensive than nozzle systems. This shouldn't be viewed as a potential power limitation. A good plate system can produce more power than the bottom ends we can typically afford. Power increases of well over 300 hp are possible on a well spec'd motor, and this can easily bring the total power of a 350 to more than 800 hp. At these levels, bottom end and block integrity are the key issues rather than the capability of the N_2O system.

Although the basic premise of an N_2O system may seem elementary, the seemingly simple process of injecting it and the extra fuel into the engine is far from guaranteeing results. For a given amount of N_2O, a poorly designed system may only produce an additional 75 horsepower, as opposed to as much as 175 hp for an optimal system. These differences are brought about by design detail. Such things as the fuel-to-N_2O ratio, fuel atomization, the effective mixing of fuel, and N_2O in the manifold and distribution are but a few of the factors that influence the production of power. These are design factors that nitrous system manufacturers should address.

A lot of experimenting with different nitrous kits was done on this 383 motor. The Nitrous Express kit shown here was their basic 50- to 200-hp plate system. During the test program a session was done with their Gemini Twin plate system and with 300-hp jets it delivered 330-hp increase. It pushed this motor to well over 800 hp. This system is one I give top marks.

During the late 1970s, a glut of nitrous system manufacturers appeared on the scene. Many failed to produce results anywhere near their claims. As a result of fierce competition, what we have left today are those companies that survived by delivering positive results. I use systems from NOS, Nitrous Express, and Zex. I've seen results from systems by these manufacturers that can be deemed satisfying by any standards.

Problem Areas

The use of nitrous isn't without potential reliability problems. Over a period of 20 years, I haven't experienced an engine failure from the use of N_2O other than experimenting on one engine to see how far I could go with Chevy's best stock parts before mechanical failure. For the record, the 350 concerned produced 1,027 hp at 6,250 rpm (863 ft-lbs) just prior to the block cracking along both banks of cylinders in the lifter valley. At lower RPM, more than 950 ft-lbs of torque was seen.

The biggest reliability problem is melted pistons, and the usual cause is inadequate fuel supply, closely followed by excessive ignition advance. Without the fuel needed for the oxygen content of the N_2O, combustion temperatures of the now lean mixture go sky high. This results in detonation and temperatures far more severe than can occur in a non-N_2O engine having too much compression and timing for the fuel octane used. The golden rule is to be sure to use the fuel pressure called for in the instructions. If the fuel pressure drops during use, problems will result. This will be your fault, not the manufacturer's.

Nitrous, if uninhibited, speeds the combustion process considerably. As a result, we find that it becomes more of a necessity to retard the ignition timing for best results as power levels increase. This can be compensated for at lower power levels, eliminating the need to spend money on an ignition retard. Here is how it's done. A typical N_2O system will have the fuel side calibrated about 40-percent rich when at the recommended pressure. This serves two important purposes: first, the extra fuel acts as a coolant to handle otherwise excessive temperatures, staving off the possibility of detonation; second, the extra fuel slows the otherwise more rapid combustion of N_2O/fuel mixes. This means that with most 100- to 150-hp plate systems, the same ignition timing can be used, whether or not the system is activated. But this situation cannot go on indefinitely. At somewhere above about 150 hp, an ignition retard of some sort will be needed.

Each manufacturer has its own pump pressure and jetting calibrations to take care of this cooling/ignition timing situation, so it's mandatory to read and comply with all the instructions before use.

System Optimization

To get the best from a nitrous system requires know-how. Starting at the cylinder and working back down the line, the first aspect to take care of is the spark plug heat range. Because most basic systems are calibrated rich, the plug will have additional cooling, but you still should check that the plugs shows no sign of overheating. Change to cooler plugs if there's any doubt concerning plug temperature.

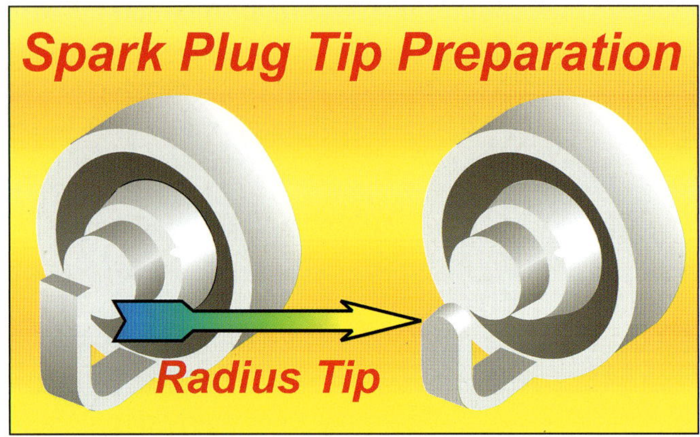

Fig 11-2. Plug prep is an important issue if you are to get the best out of your nitrous system and avoid detonation woes. Shown here is how I prep the plugs if a similarly prepped race plug is not immediately available.

This is the tip of an Autolite race plug. These work well and don't cost an arm or a leg.

If the plugs run too hot, the motor can detonate and damage will result. Sometimes, but not always, overheated plug electrodes will burn off and the cylinder stops running before piston damage occurs. This is fine as long as the plug electrodes burn faster than the pistons. You can't count on this, and it's not the recommended way to determine if the plugs are running hot. Go to a cool-running plug if you have doubts about plug temperature range. It's far easier to fix plug fouling due to running too cool than to fix pistons from running too hot!

Don't use thin-wire plugs such as the Bosch platinum, because these have inadequate heat dissipation capabilities for a N_2O-injected motor. For the record, those $3-a-pop Autolite race plugs get the job done very well and most decent-sized stores like Auto Zone have them. Failing that you can replicate the electrode form of an Autolite race plug (so long as it is of the right heat range for the job) on to a regular Autolite plug as per Fig 11-2.

Only use new and/or fully functional plug cables. Be sure the ignition system can fire the motor to well beyond the RPM required. If the ignition system drops a spark, the result can be a destructive backfire. If your HEI is equipped with an upgraded module as recommended in the ignition chapter, your ignition will be up to the job.

On the fuel side, install a pressure gauge and regulator between the pump and the nitrous solenoid. Check that the fuel pressure called for is delivered to the system throughout the full-throttle RPM range used. Using less is asking for trouble. Be sure to use adequate fuel octane. Don't use fuel that's been stored for a long time in a container that's partially full or has been vented. The light, front-end hydrocarbons that nitrous systems like probably have long since evaporated. Don't use high-octane aviation fuel. It has poor vaporization properties and doesn't work as well as automotive race fuel.

Stabilize the bottle temperature. Under most circumstances the best results are seen with bottle temperatures that deliver 900 psi of pressure. A thermostatically controlled bottle heater is a really good add-on to your N_2O installation. At 900 psi, the system doesn't deliver a significantly different amount of N_2O than at 700 psi. Although the temperature and consequently the pressure goes up, the density at the higher temperature drops and, for all practical purposes, compensates. The reasons most engines make more power at the higher bottle pressure is that mixing of the fuel and N_2O is improved, and a more favorable ratio of gaseous nitrous to liquid nitrous for better combustion is achieved.

Caribbean crew chief/racer Mervyn Bonnett ran some intake manifold tests for me to establish the sort of differences seen when the nitrous is activated. The difference between a two-plane air gap intake and a single plane race intake were virtually the same within experimental error. Manifolds with crossover heat operative showed an increase with the nitrous of about 10- to 15-hp less.

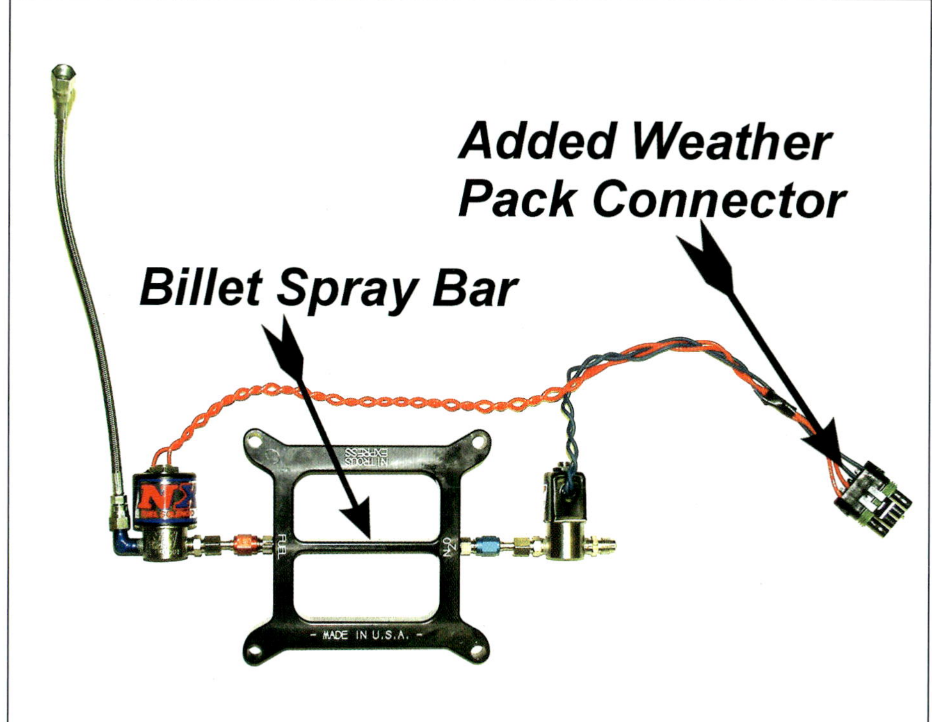

If you wire up the nitrous plate as shown here, it makes the removal and re-installing of the system fast and easy. We did this to facilitate manifold and carb swaps.

Note the annular discharge boosters on this carb. This type of booster normally delivers better fuel atomization. This works well to partially compensate for the reduced vaporization of the fuel while nitrous is being injected.

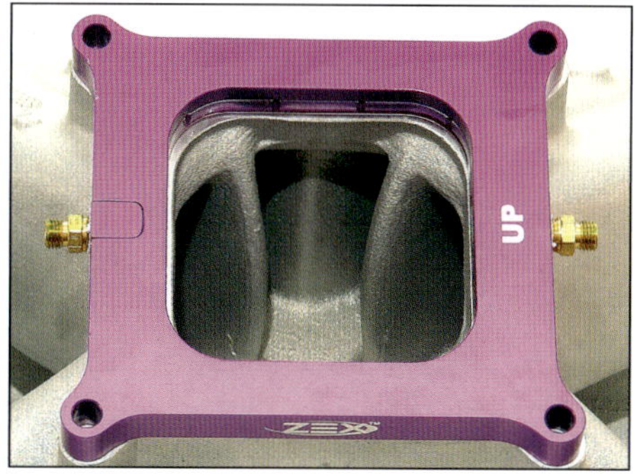

I have seen some nice results with the Zex plate system. Here I am using it to show how far off the intake opening is in terms of matching.

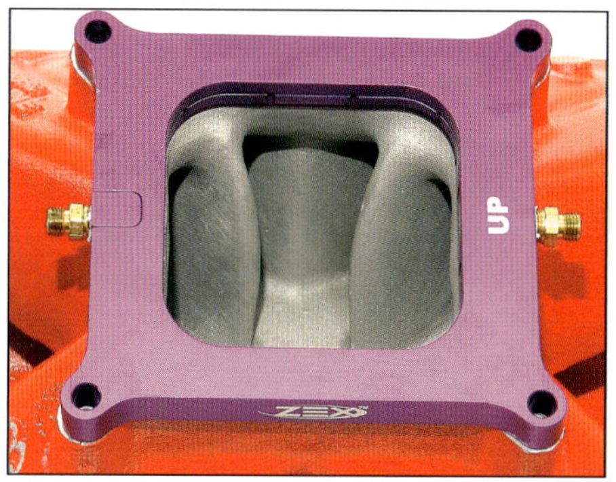

In this shot you can see how the finished intake looks. Not only has it been matched at the plenum but also at the head face. Also, the entrance to the runners has been blended. After all the internal work was done, the plenum and runners were coarse bead blasted.

More Optimal Engine Specs

To get the best from a system, we need to revise some engine specifications. First, disregard what most successful Pro Mod engine builders do, which is jack up the compression ratio to 15:1 or more. Although the dyno shows more horsepower, they are doing it wrong and for my usual consultant fee I can tell them why and what's better. For our purposes here about the best CR for both output potential and reliability with the parts we have to work with is 10:1 for a highly loaded motor (200 to 300+ hp) or 11 to 11.5:1 for a system rated in the 100-hp range. I should emphasize here that the horsepower rating of the system is only an approximation based on a typical application. If you spec your motor appropriately, you won't have a typical situation, and therefore will see more power from a given number of pounds-per-minute flow of nitrous.

For example, the basic 100-hp system from NOS normally produces between 100 to 110 hp on a typical small-block Chevy, but will produce around 180 hp on exactly the same jetting even in a budget motor if it's appropriately set up. A basic rule to follow is: Motors best respond to small amounts of N_2O with high CR, and to large amounts best with low CRs. Other than limits imposed by detonation, the significant factor affecting the best CR to use is the flow capacity of the exhaust valve.

To get the best from N_2O injection, cylinder heads need to be modified a little differently from the norm. Currently, no cylinder heads have been produced explicitly for optimizing N_2O output. When N_2O is injected into the engine, it produces an increase in exhaust volume in exactly the same manner, as would a bigger inlet valve and cylinder. Unfortunately, the exhaust flow capability of the cylinder heads generally available to us is limited. The amount of pumping losses seen on the exhaust stroke can increase considerably without additional exhaust flow capability.

A typical system delivering about 10 pounds of nitrous per minute increases the output seen in the cylinders, on the power stroke of a small-block Chevy by about 190 hp. So the now-restrictive exhaust increases the pumping losses seen on the exhaust strokes by about 80 hp. The result is a 110-hp increase at the flywheel. Retaining exhaust valve capability when such a nitrous system is used requires the exhaust valve to be increased from 1.6 inches to at least 1.8. Such an increase will effectively recover

Here some experimenting with carb CFM and valve lash is being done. For street use you can err toward the slightly small side without incurring any overall loss while possibly improving drivability. As for lash, the exhaust needed to be 0.002- to 0.003-inch tighter while the intake responded to about the same loose.

the 80 hp lost. In addition, the extra power will be achieved with a slight reduction in engine stresses. If higher-output race systems are to deliver their best, then an exhaust valve as much as 2 inches in diameter is required. It's obvious that valve increases of this order are beyond the realms of practicality, but the situation does show how important exhaust flow is for a nitrous motor. At the very least we should make the exhaust port flow as well as possible.

If the heads can accommodate it, and most stock-type heads will, the intake valve size used can be the smaller 1.94-inch item. This then leaves enough room to install a 1.7-inch exhaust valve from a Pontiac V-8 (389 to 455 ci). This move has more effect on power than you may suspect because the most critical part of ridding the cylinder of spent charge is the "blow down" phase that takes place between exhaust valve open-

ing and BDC. Because of this, the low-lift flow is more important than the high-lift flow.

The loss due to the use of a smaller intake is 10 hp at most and only comes about when the nitrous is not in use. On the positive side, when the N_2O is activated, the increase due to the bigger exhaust is about 30 hp on a typical street system.

Nitrous Cams

Although the bigger exhaust valve is the preferred method, there are limitations as to how far such a course of action can be taken. Fortunately there are other alternatives that help reduce the otherwise high exhaust-pumping losses seen when the nitrous is in use. By opening the exhaust valve earlier, the cylinder has more time to blow down. Also, the valve reaches a higher lift by the time the piston arrives at BDC. These two factors combine to make it easier for the exhaust to exit the cylinder.

In most instances, a high-lift 1.6:1 rocker on the exhaust is also of benefit, whereas it's usually not needed or is even a hindrance to power on a non-injected engine. On a 10:1 350 with a 200-hp system, opening the exhaust valve 10-degrees sooner is worth 40 to 70 hp, so we're not talking pocket change.

Checking out the specs of a functional nitrous cam may lead you to believe there's more to it than just opening the exhaust earlier. Cam companies (the ones in the know, that is) will tell you that an N_2O cam also needs to have wider lobe centerline angles and be more advanced in the engine. Although it may not be apparent at first, this is only due to the fact that the added exhaust duration is mostly on the opening side of the cam lobe. An example will clarify the situation.

Let's start with a single-pattern 270-degree duration cam for a 350 with virtually optimal timing. This will be on a 108 LCA and timed in at 4 degrees of advance. This means the intake center-line will be at 104 degrees ATDC. This gives a timing of 31-59-67-23. Now let's add 10 degrees of duration to the opening side of the exhaust lobe. Since the way the air is induced into the cylinder is virtually unaffected by the N_2O, we shouldn't alter the intake-event timing. All that needs to be catered to is the increase in exhaust volume, which is done by the earlier opening of the exhaust. Our nitrous cam will then have the valve timing at 31-59-77-23. All that's changed is the point at which the exhaust opens. In changing this, we have, by virtue of the method by which we express the LCA and the resultant advance/retard of the cam, changed the LCA to 110.5 degrees and the advance to 6.5 degrees. From this you can see that any cam that has an extended exhaust duration and is on a wider-than-normal LCA will work as a nitrous cam if it's installed in a more advanced position.

Crane had a number of cams with 110- to 114-degree LCAs that can be installed at 6 to 8 degrees of advance to good effect. This should bring their intake centerline to, or close to, the 104 required in a 350. COMP Cams has a series of five cams for nitrous applications. These have from 12- to 21-degrees more exhaust timing than intake. In a 350 they're best installed with the intake centerline at 104 degrees.

When it comes to selecting a cam, it's worth noting that a cam for a nitrous engine doesn't need to have as much intake duration as normally would have been used. Because we literally are pouring in oxygen in liquid form, the intake valve can pass all the oxygen the engine can use without going to big intake lobes. When selecting intake duration for use with nitrous, figure on about 10- to 12-degrees less than you would have picked for a non-nitrous application.

Using a nitrous-oriented cam for a motor that will have nitrous installed isn't only about getting more power. The additional 40 to 70 hp seen is a convincing argument, but there are other important advantages if your motor will power a street cruiser. Because of the wide LCA and the reduced overlap of a nitrous cam, street manners are excellent. This means that not only will you see all that extra power when the nitrous is on, but also better street drivability than the equivalent non-nitrous cam when the nitrous is not used.

A Long Arm

Although its effect is much more limited, increasing the stroke on an exhaust-limited engine (which virtually all nitrous engines are) helps extract more from the nitrous that is injected. Assuming no other changes, increasing the stroke of a 350 to make it a 383 will produce about 10-hp more from a system rated between 100 and 150 hp. Since the added stroke is usually worth about 15 to 20 hp from the naturally aspirated side, the combined effect of a longer-stroke crank means about 30-hp more without the use of more nitrous. Extra cubes from larger bores produce better results when the N_2O is in operation, but the effect isn't as noticeable as extra stroke. The main advantage of a bigger bore is that the engine makes more from its naturally aspirated charge.

Combined Effects

We've looked at the results of installing a nitrous cam, and though it gives big power increases, it's only a quick fix. The earlier opening of the exhaust valve reduces torque when the nitrous isn't in use. If this mode of operation is important, opening the exhaust valve earlier can only be taken so far. By combining a larger exhaust valve and opening the valve a little earlier, these two moves are almost additive. A system that produced an overall engine output of 640 hp was increased to 756 when the heads and cam were more appropriate for the task in hand.

High-Output Systems

As more expertise in the use of N_2O is acquired, there are benefits to fine-tuning the system and effectively injecting greater quantities of N_2O. As has been mentioned already, most systems, as calibrated by the manufacturers, are set so the fuel-to-N_2O ratio keeps piston temperatures within bounds and slows combustion so that the same timing with or without N_2O works. As greater quantities of nitrous are used, it becomes necessary to retard the ignition timing. This not only allows the engine to produce more power, but also stops it from detonating itself to pieces.

Using the spark plugs as a guide to combustion temperatures, adjust the fuel jetting and ignition timing to produce better results. A basic rule here is: Always start with the mixture a little too rich and the timing a little too retarded. Then ease up on a more optimal mixture until the plugs say "no leaner," then stop. At this point, go back to one jet-size larger on the fuel, and then turn your attention to the timing. Add in no more than 2 degrees at a time, and as soon as worthwhile gains are no longer seen, stop. Stopping just short of optimum will allow a useful buffer zone to accommodate those times when things get a little hotter than anticipated.

Nitrous power increases above 200 hp can become difficult to put to the ground. This is a nice problem to have, but it won't win races. When the higher levels of nitrous-augmented horsepower are required, it's worthwhile going to a two-stage system. Two-stage plate systems aren't significantly more expensive than a single-stage system. Instead of overwhelming traction with excessive low-end torque, the amount of nitrous at low RPM can be reduced, and as RPM increases, the second stage can be activated. Not only does this make the use of nitrous more manageable off the line, it also allows a greater quantity of nitrous to be used as the amount the engine will tolerate increases with RPM.

Because the rate of N_2O injection is independent of RPM, we find, if we ignore internal engine friction for a moment, that the amount of additional horsepower produced is about the same regardless of RPM. This means that a system producing a 150-hp increase results in a torque increase of 315 ft-lbs at 2,500 rpm. At 4,000 rpm, 150 hp equates to 197 ft-lbs. At 6,000 rpm, that same 150 hp is only 131 ft-lbs. The increase in cylinder pressures obviously follows suit. Dumping in too much N_2O at low RPM can produce gasket-blowing cylinder pressures. Because of the high cylinder pressures at low RPM, it isn't advisable to use anything other than a moderate system at revs below 3,000.

This big increase in low-RPM torque means that faster acceleration and quarter-mile times are achieved by using fewer RPM than if the nitrous isn't used. It also means that the amount of additional stall required in the converter is less. High-stall converters are made for engines that don't produce any power until they're up on the cam and revving well into their powerband. If a nitrous motor is backed up by a converter with too little stall, it will be compensated for by the nitrous producing more low-end torque.

Because a tighter converter is more efficient, it allows a nitrous car to run significantly better MPH than if a looser converter were used. In many applications a converter no looser than a typical street unit is required. However, if a lot of torque is fed into a stock converter, it's likely to balloon and/or break the fins. A heavy-duty converter will be required with any package making more than 400 total horsepower.

As the torque produced by the nitrous drops off at higher RPM, the engine can tolerate the use of more nitrous. By two-staging, we make the system more chassis friendly and also make more top-end power.

Although it may fall outside your budget, there are some tools that can help make a more sophisticated and effective nitrous system. MSD, Accel, and Holley all produce an ignition package that allows the use of a fixed distributor with the advance curve generated electronically. All these units allow the triggering of the nitrous to retard the spark timing. Also, they allow for a nitrous shutoff a few RPM before the rev limiter is reached. In addition, multiple rev limiters can be used so that one limit is used for the burnout, another for launching, and the final one for high-end limit. All these units make for better use of the animal-like output that nitrous generates. Since there's a relatively wide spread in cost, I suggest that you get a catalog from each of these companies, then determine which unit best meets your financial needs.

How Much Power?

There are two ways to use nitrous. It can be used to augment an engine that's largely optimized for use without the nitrous. This means the nitrous power is simply an add-on and will give a good increase but not necessarily to its full potential. The engine in this instance could well be described as a nitrous-assisted engine. The fact that the nitrous isn't giving its best is partially offset by the fact the engine, without it, is. However let's strike a middle-of-the-road deal here and have the cam halfway biased toward nitrous use. In this instance the cam is an XR282HR COMP hydraulic roller (this is a post-1987 hydraulic roller block). The cam at 230/236 at 0.050 is relatively mild and ideal for street/strip use with a strong bias toward street use. This cam on a 110 LCA was set in at 6-degrees advanced. The chart on the left on page 149 shows the output of this motor with out-of-the box EQ heads and Zex's cheapest perimeter spray plate system.

The jets for this unit are calibrated to produce a nominal 100, 125, 150, and 175 hp. In this instance I opted for the 150-hp jetting. At a total cost of under $5,300 as a turnkey deal (includes carb and full exhaust), this motor was a true budget street performer. A smooth 600-rpm idle and perfect road manners belied its ability to rip off mid-11-second quarter-miles in a 1970 Camaro.

If we're building for maximum power with the nitrous in operation and leaving the power to fall where it may, some figures are possible that are limited more by component strength than by power production capability. For a drag-race application where finances dictate that the engine lasts at least 100 passes, it's best to limit power to about 850. The chart below right shows the results of an engine that was conservatively jetted with the idea of allowing it to survive for a couple of season's worth of racing. This engine made only a relatively modest 440 hp without the nitrous, but that went to 780 with the single-stage system activated.

This engine was a parts-collecting exercise if there ever was one. A new 750 mechanical-secondary Holley sat on top of a well-used swap-meet Victor Jr. intake. The heads (Dart) were extensively ported and utilized the usual 2.02/1.6 valve combination. With the flat-top Ross pistons, a 10.5:1 CR was achieved. A set of Isky 1.6:1 rockers, found at a swap meet, were actuated by a roller cam. This was acquired as a damaged part from a motor that suffered a rod failure due to excessive RPM. One cam intake lobe and three roller lifters were damaged. This package cost $30. The exhaust profile was 305 seat duration, ideal for the nitrous application. The cam was sent back to the company that originally ground it and a much shorter 284-degree profile was ground on the intake lobes. This fixed the damage and allowed the LCA to be widened to 112 degrees. For about $160 we had a roller cam and a set of lifters. This setup didn't require heavy-duty valvesprings because RPM involved weren't that high. The crank was a stock GM forging ground 20 under and the rods were a set of Crower heavy-duty 5.7-inch long items that we had owned forever. A prepped stock pump and a budget Moroso oil pan took care of the bottom end. Ignition was by means of a hopped-up HEI with 28-degrees total when the N2O was in operation. Headers were 1.75-inch-diameter and brand new from Walker. A 15-inch collector and open exhaust were used. Fuel was courtesy of Unocal. This motor was a lot of work, but it cost only $2,900.

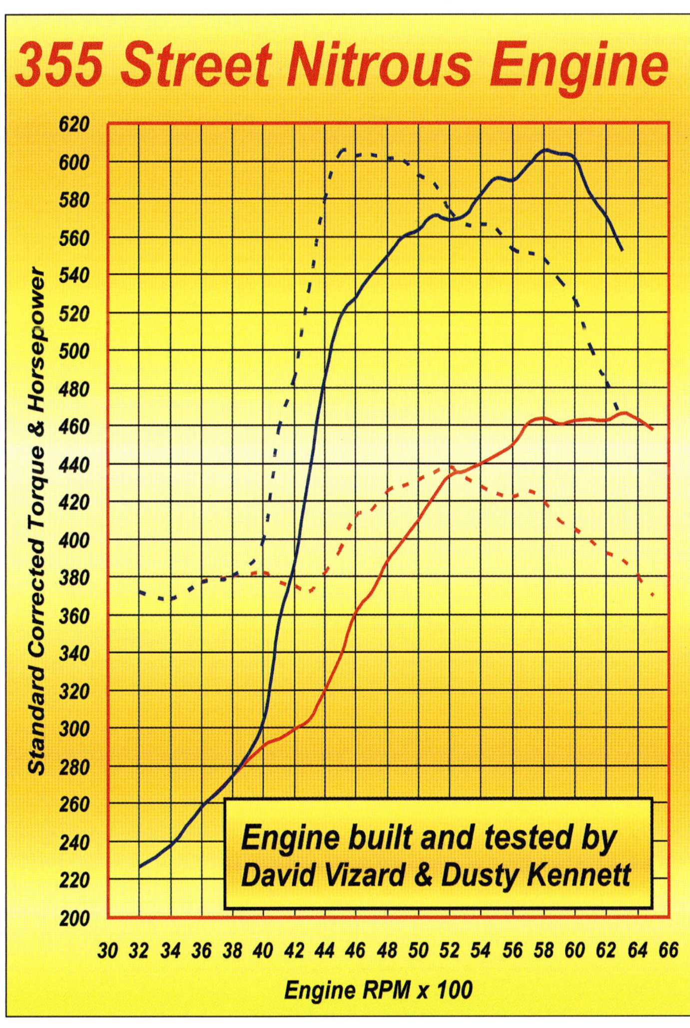

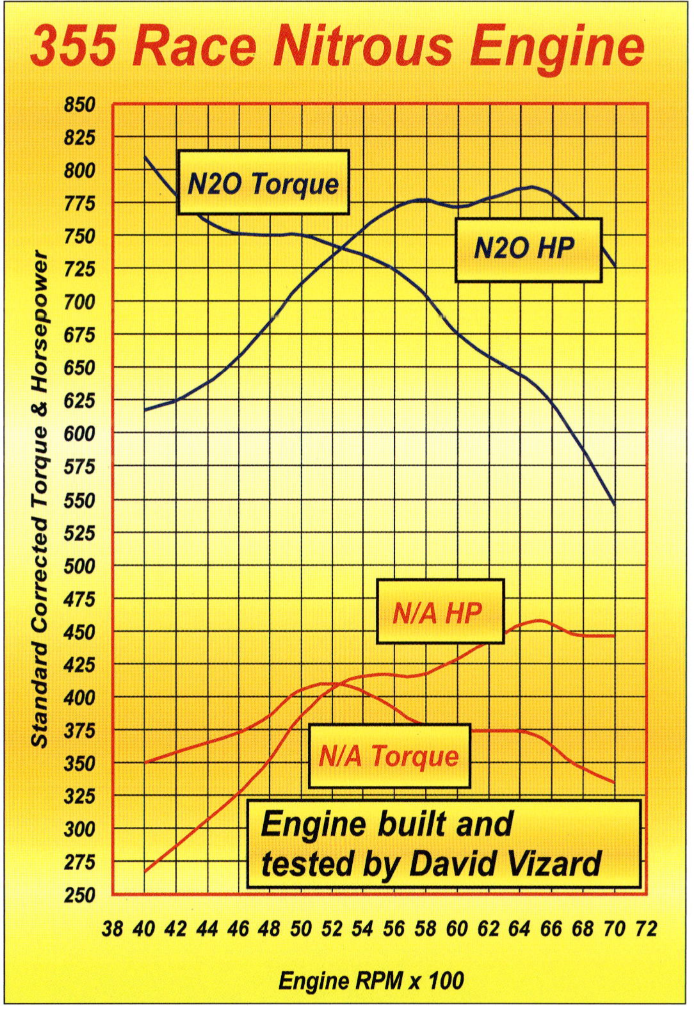

TEN HOT ENGINE BUILDS

It seems that I am far from the only one on a quest for best power per dollar spent while retaining reliability. Shown here is an engine from Blueprint Engines that seeks to do the same. I had to build, in my shop/studio, a replica of one of their engines from a kit they sent me for a magazine article. The result, from their baseline engine for truck usage, was 453 horsepower and 495 lbs-ft from 383 inches. All that came assembled at a turn key cost of $7,200 including a 50,000-mile warranty!

Since the first edition of this book, we have suffered the inevitable ravages of inflation. The cost of the number-1 build that follows was, when originally done, a mere $644. That was around 1981. Since then inflation has devalued the dollar some 43 percent so, on that build, everything will, in the main, cost about 75-percent more. This brings the cost to $1,127. As you can see, all the costs stated in the first edition will be out by quite a factor. To compensate here I am going to do my level best to go through the components used in whatever early builds are still included here and estimate costs at today's prices.

A note of caution here: Remember that the use of a flat tappet cam will mean taking care of the lubrication requirements of the profile/lifter interface. Since the removal of ZDDP from the formulation of most oils (2006), be sure to use an oil that does cater to the needs of a flat tappet cam or use an additive that does so. Just as a refresher, read what I have to say in the cam chapter on this important subject.

Engine Number 1: Lowest of the Low

I built this motor in 1981 to test my then-new dyno installation. By carefully inspecting the parts to make sure they were fit for further use, a super cost-effective motor was achieved. This engine could be construed as a reconditioner's nightmare since too many like this would cause a lot of business failures. Only two machining operations were done on the block. The block was decked to 8.990 and the bores were honed. The small piston protrusion produced allowed the crowns of the cast pistons to be machined to give precisely −0.005-deck height. The piston skirts were knurled and were file fitted, using paper as a feeler gauge, to give 0.003-inch clearance. The cam was a respectable one dug up from a number of cores, and the lifters reconditioned by rubbing them in a figure-eight pattern on 80-grit emery until a new working surface was generated. They then were stripped and cleaned. The intake manifold was mildly modified to the extent that the plenum-to-runner junctions were rounded off and the ports cleaned up and port matched to a Fel-Pro 1204 gasket. The heads were given a basic pocket port job after being machined to take 0.003-inch oversize stem from 2.02/1.6 valves. As to how much more power this made than an absolutely stock motor is difficult to say with any degree of precision. The nearest I could find was a stock 1980 Z-28 motor with the smog exhaust system replaced with a set of 1⅝-inch headers and some turbo mufflers. This factory stock engine made 234 hp and 321 ft-lbs of torque.

As you can see, this was a bargain-basement deal. It made over 300 hp for a total, in 2008 money, of $1,127. Unfortunately, this motor never went in a vehicle. It spent about six weeks on the dyno, where it ran flawlessly in testing various aspects of our installation for future exhaust system noise and performance testing. During the period it was on the dyno it made about 700 to 800 pulls, which I would estimate was equivalent to doing a 400-mile race with a 5,500-rpm limit. After this extended session, the motor came off the dyno and was used over a period of several years as a "stock" mule motor. After sterling service, it was torn down and showed no signs of undue wear, suggesting, with quality oils, a life of at least 100,000 miles.

Core: 350-ci engine circa 1970–1971. From crashed car at wrecking yard. Estimated mileage: 150,000 well-maintained miles. 2008 cost: $306.

Block: Honed to clean up at +0.005-inch oversize. Decked to 8.990 so at stock compression height piston will be- a few thousandths out of block.

Pistons: Original pistons knurled and file fitted to bores with approximately 0.003 clearance. Machined crowns in lathe to put them 0.005 to 0.007 below block deck face. If the amount of piston rock is an unknown factor, recommend 0.010 down bore for piston crowns.

Final CR: 10.1:1

Crankshaft: Stock cast crank polished. This came out to bottom limit to a couple of tenths below bottom limit.

Rods: Stock but mildly lightened and rod fixture balanced.

Oil Pump: Stock but detailed.

Pan: Stock

Cam: A good used stock Chevy 929 advanced 4 degrees. Timing set, budget stock replacement.

Lifters: Reclaimed

Valvetrain: Stock springs shimmed 0.030 tighter, reconditioned rockers, stock pushrods, 0.080 lash caps.

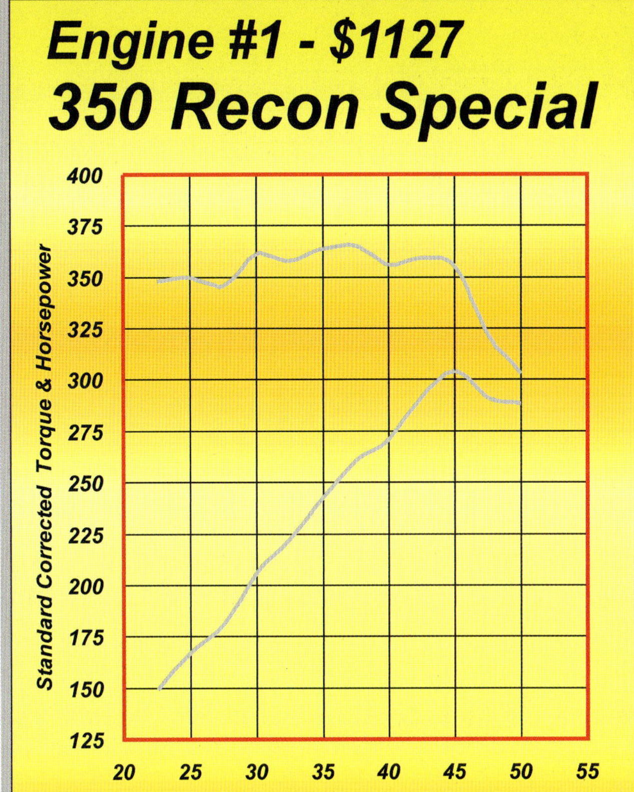

Heads: 041 castings with new oversize-stem (+0.003) 2.02/1.6 valves installed. Casting then pocket ported and flat milled to 64 cc.

Intake: Mildly reworked stock iron Q-Jet manifold.

Carb: Q-Jet with throttle shafts thinned down.

Air filter: K&N

Ignition: Stock HEI

Plugs/cables: Autolite/Pep Boys brand.

Headers: Walker 1⅝ inch.

Mufflers: Dynomax 2½-inch Turbo.

Parts: $590
Labor: $537
Total: $1127

This is the baseline motor and the curves will be shown in light gray on almost all the subsequent graphs for comparison. It's not being used as a baseline because it is the lowest in output but because it's the lowest cost. All things considered, this motor does well on output primarily due to the pocket porting of the heads and the 10.1:1 CR.

Engine Number 2: Basic But Full Recondition

This motor was about as basic as you can get for a stock rebuild. It went into my 1974 Chevy truck and, with all the emissions hooked up, idled like a watch at 600 rpm and ran a whisper-quiet 14.9-second quarter at 93 mph. The spec was simple. A 650 spread bore Holley was used on a mildly ported production iron manifold with the heat riser blocked off. From here, the cooler, denser charge passed through the pocket-ported heads into polished chambers. The pistons were the popular D-shaped dish style, which seem to produce a highly effective chamber shape (with the heads recommended) for engines requiring a compression of 9:1 or less and using regular fuel. This particular engine ended up at 8.6:1, but experience indicated that 9:1 could have been used and would be okay on current formulations of regular 87-octane fuel.

Exhaust was via a set of Walker 1⅝-inch truck headers that dumped through a set of high-flow Cyclone Hemi mufflers. The present day equivalent to these mufflers would be the Walker Dyno-Max series design. Prior to being installed in the vehicle, this motor was a dyno mule for about two months. During that session, it was run with a COMP Cams 270H on a 108 LCA. This cam is well suited to motors with 8.5 to 9.5:1 CR and, as you can see from the graph, bumped the horsepower by almost 50 or so, as well as delivering a big increase in torque. If you want to spend the extra $250, this is what you'll get. As installed, the fuel mileage was good at about 15.5 combined city/freeway and about 19 freeway. A lockup converter and a later four-speed auto would have helped the mileage considerably. I'd like to have reported how this motor fared over an extended period, but after I had about 20,000 on the motor, the truck was stolen.

Core: 350-ci engine circa 1971–1972. Obtained from local engine reconditioning shop. 2008 cost: $360.

Block: 2-bolt block bored +0.030 and decked to 9.000 inches.

Pistons: Speed Pro with D shape dished crown or KB. 118 hypereutectic 22-cc dish pistons.

Final CR: 8.6:1

Crankshaft: Stock-size journal cast crank polished.

Rods: Stock but mildly lightened and rod fixture balanced.

Oil Pump: Stock but detailed.

Pan: Stock

Cam: COMP Cams 270 H on a 108 LCA in at 4 advance. Timing set: budget stock replacement.

Lifters: Stock Sealed Power

Valvetrain: New stock springs shimmed 0.030 tighter, new rockers, stock pushrods, 0.080 lash caps.

Heads: 186 castings with new bronze guides new 2.02/1.6 valves installed. Castings pocket ported and flat milled to 64 cc.

Intake: Mildly reworked stock iron Q-Jet manifold.

Carb: Holley 650 Spread Bore.

Air filter: K&N

Ignition: Stock HEI.

Plugs/cables: Autolite/Accel

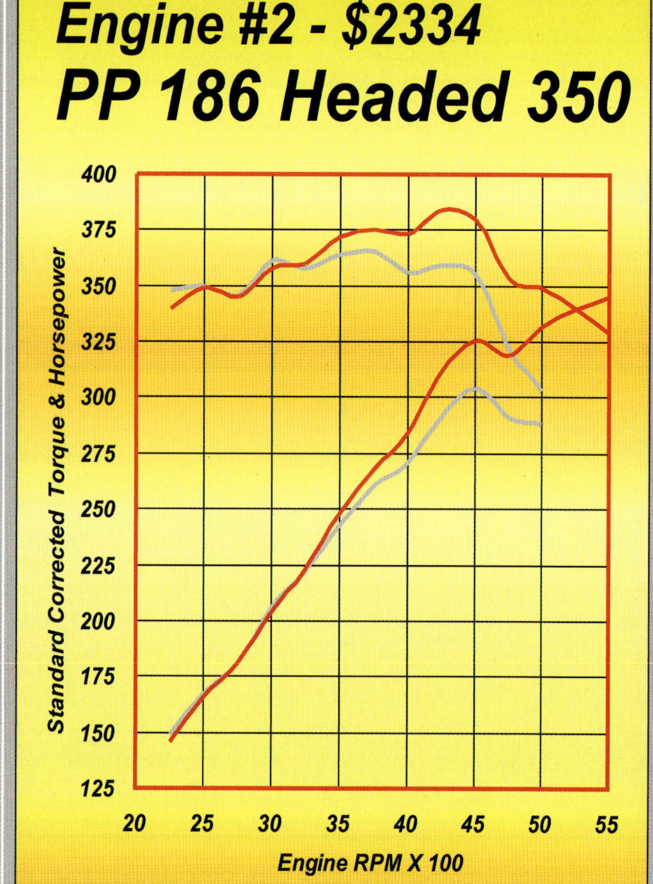

Headers: Walker 1⅝-inch.

Mufflers: Dynomax 2½-inch.

Parts: $1797
Labor: $537
Total: $2334

For all practical purposes this is engine number-1 but with a few refinements and D-dish pistons to drop the CR from 10.1:1 to 8.7:1 for use with 87-octane fuel. The COMP Cams 270H cam and the reduced compression combined to drop torque under the 2,500-rpm mark, but from there up the slightly longer duration and significantly higher lift started to pay off. Peak torque rose over our baseline engine by almost 20 ft-lbs and peak power went up by about 50 hp. With a smaller dish as per the 12-cc KBs (for a 9.4:1 CR) this spec will top the 360-hp mark with ease and re-gain the torque below 2,500 rpm. Expect to use 89-octane fuel if you go this higher CR route.

Engine Number 3: Basic Street Mods

For the guy that is willing to do some semi-serious porting, this motor represents a basic, sensibly hopped up street motor that's totally practical as a day-to-day driver. This build will return decent fuel mileage and propelled a 3,400-pound car, with mufflers and on street tires, down the strip in the low 13 seconds at about 105 mph in the lights. Unless the tires are good figure you will make a lot of smoke.

The carb on this motor was a 750 mechanical secondary reworked as described in the induction chapter to produce about 820 cfm. This, together with a 1¼-inch open spacer, was mounted on a port- and plenum-matched Edelbrock Victor Jr. I did this build well over 12 years ago but if I had known as much about making good top end from an air gap two plane, that is what I would have used. Experience tells me this would produce far more low speed torque with, at this power level, only a minimal drop in top-end output. The basic full porting job on the 186 heads produced 235 cfm at 0.600-inch lift on the intake and 200 cfm on the exhaust. If I were building this engine today I would opt for a set of performance iron heads from EQ, Dart, or RHS. This would get better results without having to spend about 50 hours porting. With 10.1:1 compression, the stock HEI was running out of spark power. This was fixed with a Performance Distributors high output module. The rest of this engine was as listed in the preceding chart. A higher output than this can be expected if the catalytic converter-equipped exhaust system used for this motor is replaced with duals. It is important to understand the catalytic converters are not created equal. If you have to use them I can vouch for the Walker hi-flow items. Anything less than a good, solid 400 cfm will cut the output of any performance build considerably.

This build combination cranked out 434 hp and 428 ft-lbs of torque. Idle was smooth 750 rpm and was virtually lope free. The Walker mufflers used cut the noise but left a muted yet intimidating exhaust note.

Core: 350-ci engine circa 1971–72. Obtained from local engine reconditioning shop. 2008 cost: $360.

Block: 4-bolt (but 2-bolt would also be satisfactory) block bored +30 and decked to 9.000 inches.

Pistons: Speed-Pro or KB flat-top hypereutectic.

Final CR: 10.2:1

Crankshaft: Stock-size journal forged crank polished.

Rods: Stock lightened and rod fixture balanced.

Oil Pump: Reworked stock.

Pan: Moroso basic street.

Cam: COMP Cams 268 H on a 108 LCA in at 4 advance. (for optional and better output use a single-pattern 268 Xtreme energy grind number 5443). Timing set, budget stock replacement.

Lifters: Stock Sealed Power

Valvetrain: COMP 981-16 springs, new Magnum roller tip 1/6/1 COMP rockers, COMP magnum pushrods.

Heads: 186 castings fully ported with new bronze guides and 2.02/1.6 valves installed. Castings flat milled to 64 cc. Recommended alternatives: as cast heads from EQ, Dart. or RHS.

Intake: Edelbrock Victor Jr with 1¼-inch open spacer.

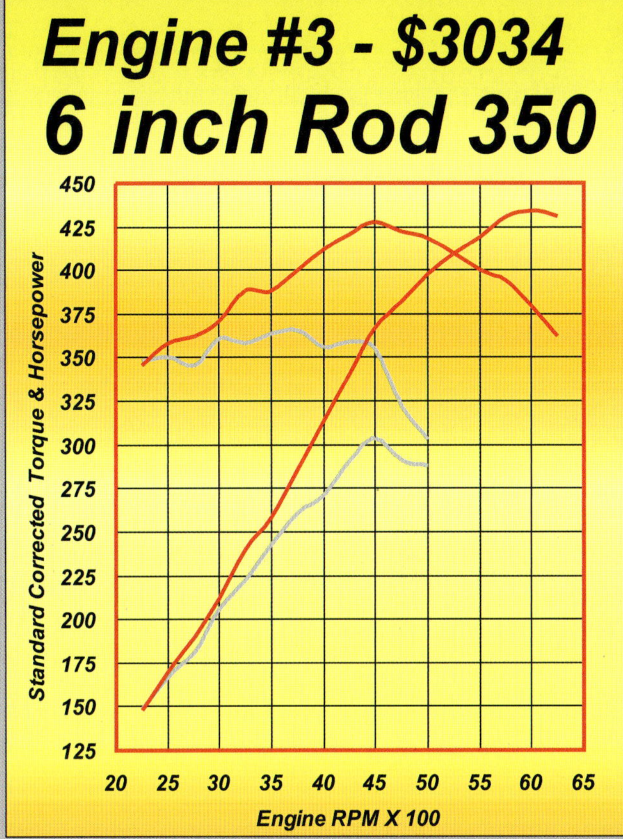

Carb: Holley 750 with shaft and butterfly mods plus choke horn removed and blended out. Final flow: 820 cfm.

Air filter: K&N

Ignition: Stock HEI with Performance Distributors module.

Plugs/cables: Autolite/Moroso

Headers: Walker 1⅝-inch.

Mufflers: Dynomax 2½-inch.

Parts: $2388
Labor: $646
Total: $3034

This build was very successful for its day, but the big drawback would be the fact that there is 50 hours of head porting involved. If built with an out-of-the-box iron casting from any one of the three alternatives mentioned, similar if not slightly better results can be obtained for far less build hassle. If any one of these alternative head casting are used, figure a cost rise of about $350 to $400 on that quoted here.

Engine Number 4: Roller Blocked 350

GM has produced millions of roller blocked small-block Chevys since 1987. Since about 2006, these blocks are all I have worked with. If you buy yourself a good used core from a fuel injected vehicle you can get some dynamite results in terms of torque, power, and fuel efficiency in a very cost-effective manner by following this build. First, the bore wear is usually minimal to near zero so a quick pass with a hone and you're in business. The pistons can be re-ringed and are ready for another 150,000 miles if they are the forged dished type. This is what I did for this particular build—that, and swap out the stock roller cam for a custom single-pattern COMP Magnum hydraulic roller grind number 3119. This cam, ground on a 108 lobe centerline angle (very important to stick that LCA), with an advertised duration of 280 degrees, had 224-degrees duration at 0.050 lift delivered; with 1.6 COMP Magnum rockers, 0.560 lift. Springs used were COMP 26918 beehives. This together with a Professional Products Crosswind two plane and a manual choke 750 Holley and pocket ported Vortec heads (you can substitute a pair of Engine Quest CH350C heads here for better results yet, and build is priced out for such) delivered a stout 435 hp and 437 ft-lbs of torque. The 10:1 compression ratio, achieved by milling the head 0.025, required the use of premium fuel but the resultant fuel mileage from the high compression was good, so cost per mile was acceptable.

Ignition for this was handled by a custom-curved Performance Distributors HEI with vacuum advance. That in itself was a great contributor toward the decent street mileage this engine produced. As for cost, a lot of new parts went into this build such as new valve covers, oil pan, etc. If you have these parts or get used ones, about $300 can be saved on this build.

Core: 350-ci engine circa 1993. Obtained from AAEQ. 2008 cost with shipping: $660.

Block: 4-bolt (but 2-bolt would also be satisfactory) 880 block honed 0.001 over to clean up and decked to 9.000 inches.

Pistons: Stock re-ringed.

Final CR: 9.5:1

Crankshaft: Stock-size journal forged crank polished.

Rods: Stock

Oil Pump: Reworked stock.

Pan: Moroso basic street.

Cam: COMP Cams 280 Magnum on a 108 LCA in at 4 advance. Timing set, budget stock replacement.

Lifters: Stock factory roller lifters.

Valvetrain: COMP 26918 beehive springs, new Magnum roller tip 1/6/1 COMP rockers, COMP magnum pushrods.

Heads: As new Vortec castings pocket ported, 2.20/1.6 valves and flat milled to 60 cc Recommended alternatives: as-cast heads from EQ, Dart, or RHS.

Intake: Professional Products two plane Crosswind.

Carb: AED 750 Holley

Air filter: K&N

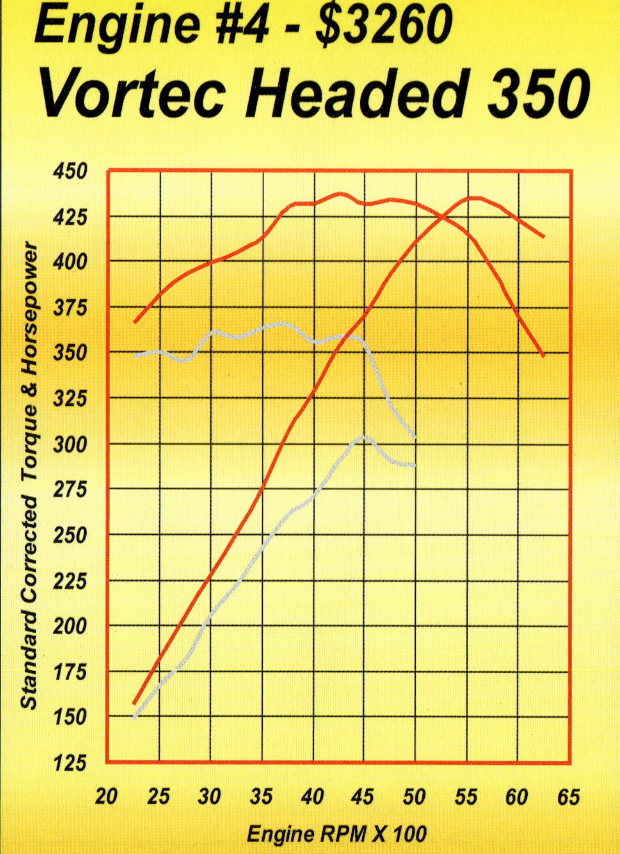

Ignition: Stock HEI with Performance Distributors module.

Plugs/cables: Autolite/Accel

Headers: Walker 1⅝-inch.

Mufflers: Dynomax 2½-inch.

Parts:	$2580
Labor:	$190
Total:	$2760 with Vortec heads
	$3260 with EQ heads

Going with the later hydraulic roller cam block is the most cost-effective way to get into a roller cammed motor. The advantages are not just the fact that the block comes stock with a roller, but also rods, pistons, and a block of better quality than the pre-1987 engines. Here, a totally streetable 437-hp engine was built that, given top grade oil, can go 100,000 miles without wearing itself out. Also, the 9.5:1 CR means it will run on the mid-grade 89-octane fuel.

Engine Number 5: Zex Nitrous 350

If you are looking for big numbers and a quick build time to go fast at the strip on not too much money, here is the motor for you. The core was a 1993 350 in near-complete form that I got for $500. Here is how this originally injected engine finished up. Starting at the top of the engine was an AED prepped 750 Holley costing about the same (sometimes less) than the Holley you can pick up at your local speed shop. This one, though, is custom calibrated by experts just for your spec motor. This sits on a Zex perimeter nitrous plate, which in turn is on a port and plenum matched Edelbrock Super Victor. Heads were a set of EQ 23s in out-of-the-box form. The 64-cc chambers, with the flat-top KB pistons delivered a 10.2:1 CR. The block was, as mentioned, a late model 880 roller cam block so it made sense to stick with a hydraulic roller valvetrain and use the stock rollers. I had COMP grind, to my specs, a special nitrous grind that used an Xtreme Energy profile number 3196 on the intake and 3318 on the exhaust (288/300). This was ground on a 110 LCA and installed 6-degrees advanced (intake centerline at 104). A stock crank and the budget Scat Premium 7/16s thru-bolt 6-inch rods I like so much were paired off in an otherwise stock bottom end. The results with 200-hp jets in the Zex system were very invigorating with this low-cost engine delivering 677 hp.

Core: 350-ci engine circa 1993. Obtained from AAEQ. 2008 cost with shipping: $560.

Block: 4-bolt block honed 0.001 over to clean up and decked to 9.000 inches.

Pistons: KB 702 flat-top forged stock size.

Final CR: 10.1:1

Crankshaft: Stock-size journal cast crank polished.

Rods: Scat Premium 7/16 cap bolt.

Oil Pump: Reworked stock.

Pan: Stock

Cam: COMP Cams 288/300 on a 110 LCA in at 6 advance. Timing set, budget stock replacement.

Lifters: Stock hydraulic rollers.

Valvetrain: COMP 26918 beehive springs, Aluminum roller rockers 1.6 intake and 1.5 exhaust. COMP magnum pushrods.

Heads: As-cast EQ23 heads.

Intake: Zex perimeter nitrous plate on Edelbrock Super Victor with 1-inch open spacer.

Carb: AED 750 Holley with shaft and butterfly mods plus choke horn removed and blended out. Final flow: 815 cfm.

Air filter: K&N

Ignition: Stock HEI with Performance Distributors module.

Plugs/cables: Autolite/Accel

Headers: Walker 1⅝-inch.

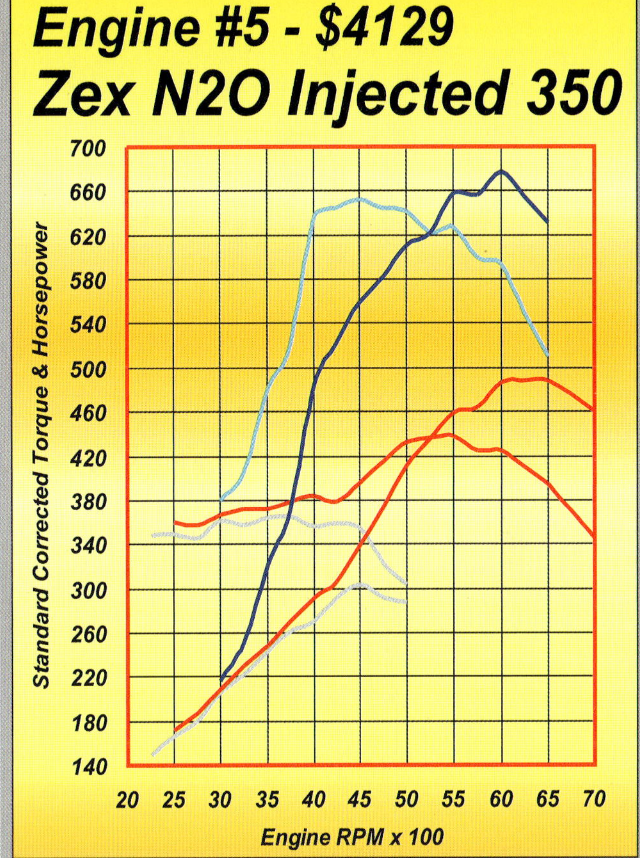

Mufflers: Dynomax 2½-inch.

Parts: $3939
Labor: $190
Total: $4129

If driving to work at a leisurely pace and going fast on the strip on the weekends fits your needs, then this build should be seriously considered. With a manual trans, gas mileage is better than you might otherwise expect from a motor that makes about 480 non-injected horsepower. By not purging the system overly, the nitrous can be made to hit softer, thus reducing the tendency to spin the tires after a clean launch. The torque curve (light blue) shows how, when activated at 3,000 rpm the system comes on progressively until it is at 100 percent at 4,000 rpm.

Engine # 6: Big Power Small Budget 350 N₂O Motor

This was an interesting motor, with the parts coming together over a period of time. Essentially it started with the acquisition of a good block and a set of used Crower heavy-duty rods from a swap meet. Most Chevy small-block race engines use 6-inch-or-longer rods, and the 5.7 Crower rods located at the swap meet were the victims of progress. However, for our purposes, the fact that these rods were marginally less than ideal length was more than offset by their substantial strength. For pistons, I used a set of off-the-shelf Ross forged items. This company makes a nice piston at a very acceptable price. Although lighter than stock, the combination of the heavier Crower rods and the Ross pistons meant a difficult balance job. Here I eliminated the need for a heavy metal balance job by putting a 5/8-inch drill through the cheeks of the big ends of the factory forged crank to make them hollow. This more than countered the additional weight of the 720-gram Crower rods, compared to the 600-gram stockers.

For the first time around I built this engine with heavily ported 041 factory castings and used 1.94 intakes and 1.65 exhausts. This worked very well but shortly after about 90 passes, the heads started cracking so this engine was put into storage for a while pending new head casting. A year or two later a set of sound Bow Tie iron heads came into my possession really cheap. It seemed that someone had attempted porting them and had made poor progress. They looked a disaster but I could see that they were largely redeemable. After porting and a seat job these were then ready for use on the engine.

For an intake manifold a Victor Jr. was used along with about the cheapest plate nitrous system offered by NOS. The ignition system was a $10 salvage-yard unit reconditioned and equipped with an MSD module.

On the original ported 041 heads this motor made, as a single stage injected unit, some 775 hp on the plate system. After the second build with the Bow Tie heads and a little more nitrous, the single stage output, with supposedly 300-hp jets, went up to almost 800 hp.

The biggest problem with this build even in its initial form was that it had a propensity for breaking traction. The fix was a higher rear end ratio and a second stage was added to make more top-end speed. What I did here is plumb in a port injection system and use it for the first stage. The original plate was calibrated with 100-hp jets and was used as the second stage. Power on the port injected first stage was 790 hp. The system was triggered at about 3800 rpm and consistently showed 800+ ft-lbs at 4,000 rpm. At 5,200 rpm, the secondary (plate system) came into action and boosted the output to 885 hp at 6,500 rpm. Torque, with the second stage operative, was 825 ft-lbs at 5,500 rpm.

Although the results achieved with the two-stage system were well worth the expenditure, the total costs did go up a fair bit. The cost of the motor as per the single stage setup is what is shown in the nearby graph. To build the two-stage system with the electronics to run it (an Accel three-stage rev limiter and window switch) would, in today's money, add a little over $1,400 to that cost.

Core: 350-ci engine circa 1993. Obtained from AAEQ. 2008 cost with shipping: $560.

Block: 4-bolt block bored +0.030 over and decked to 9.000 inches.

Pistons: Ross flat-top forged 0.030 oversize.

Final CR: 10.1:1

Crankshaft: 0.010 undersize-journal forged factory crank.

Rods: Crower 5.7 long billet 7/16 cap bolt.

Oil Pump: Reworked stock.

Pan: Stock

Cam: COMP Cams Xtreme Energy solid roller 280/292 (Profile number's 4874/4876) on a 110 LCA in at 6 advance. Timing set, COMP Cams adjustable.

Lifters: COMP 818-16 solid rollers.

Valvetrain: COMP 987 springs, Aluminum roller rockers 1.6 intake and 1.5 exhaust. COMP magnum pushrods.

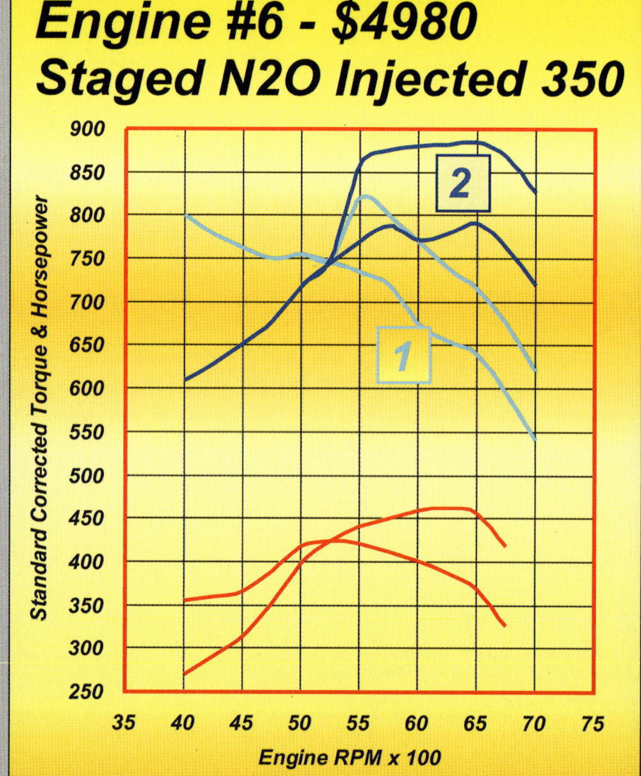

Heads: Bow Tie iron heads.

Intake: NOS nitrous plate on Edelbrock Super Victor plus NOS port injection.

Carb: AED 950 series Holley with shaft and butterfly mods plus choke horn removed and blended out. Final flow: 945 cfm.

Air filter: K&N

Ignition: Performance Distributors HEI

Plugs/cables: Autolite/Accel

Headers: Walker 1⅞ inch.

Mufflers: N/A

Parts: $4340
Labor: $640
Total: $4980

If the parts involved are strong enough, then Nitrous is the way to go for cheap power. Using just the NOS plate system with 300-hp jets this engine made almost 800 hp. By adding a port injection system and using it as the first stage and the plate as the second stage the output shown here was seen. The light blue curves (number 1) are torque for the engine during nitrous operation. The second higher torque curve shows how the second stage plate system added to the power when triggered at 5,200 rpm. The dark blue curves (number 2) show the power output in both single and two-stage form.

When the nitrous is not in use this engine drives almost like a stocker and could be used on the street without any drivability issues. Low speed torque however, because of the nitrous cam and the big headers was not the best but acceptable to the die-hard go-fast enthusiast.

Engine # 7: Entry-Level 383—Big Torque, Small Cost

This build is a cost-effective cross between street functionality and the drag strip. It is more than civilized enough for day-to-day driving, even if used in a working truck, yet makes enough grunt to show well at the drag strip—even in a truck! It was one of the last builds I did with ported factory castings. These are scarce now so although I show the power curves with ported 186 heads you will actually get better results with 180-cc port aftermarket iron heads from EQ, Dart or RHS. The cost of going this route is not that much more because the heads I did required new guides, studs and guide plates, a seat job etc. new castings won't have these costs. In conjunction with a set of flat-top KB pistons, the heads delivered a 10:1 CR.

Also, this motor used Scat's cheapest 3.75 stroker crank and 6-inch rods and that is how I priced it out here. These rods meant time-consuming hand clearancing of the block, which, if you want to avoid, means spending about $80 more on Scat's stroker rod. If you factor that and the cost of a set of aftermarket heads into the deal this build will run about $350 more than shown here.

For a cam, this motor used a COMP Xtreme Energy, single-pattern flat tappet hydraulic grind with the 270-duration (advertised) lobe number 5444, on at a 106 LCA 4-degrees advanced. Also, be sure when you order the flat tappet cam from COMP you have it nitride hardened; otherwise, you could be replacing it real soon because of the removal of ZDDP from modern oils.

Induction was by means of an Edelbrock Performer RPM Air Gap and an AED 750 vacuum secondary Holley 4150-style carb. However, an Edelbrock Performer with an 800-cfm Edelbrock carb looks like it might have the edge when it comes to city-driving mileage.

For ignition I used a stock HEI that I rebuilt using a Performance Distributors module and a new cap and rotor.

The power curves shown here are as I built the engine but the upgrades (heads, etc.) mentioned will improve torque by maybe about 10 ft-lbs but the horsepower can go up by as much as 20.

Core: 350-ci engine circa 1980. Obtained from AAEQ. 2008 cost with shipping: $530.

Block: Four bolt block bored +0.030 over and decked to 9.000 inches.

Pistons: KB cast hypereutectic flat-top 0.030 oversize.

Final CR: 10:1

Crankshaft: Scat 3.75 9000 series cast steel stroker crank.

Rods: Scat 6-inch through bolt (recommend Scat's stroker rods as better alternative).

Oil Pump: Reworked stock.

Pan: Stock

Cam: COMP Cams Xtreme Energy Hydraulic flat tappet single- pattern 270 (Profile number 5444) on a 106 LCA in at 4 advance. Timing set, COMP Cams 3200.

Lifters: COMP 812-16 hydraulic flat tappet.

Valvetrain: COMP 986 springs, Aluminum roller rockers 1.6 intake and exhaust. COMP Magnum pushrods.

Heads: Fully ported 186 iron heads with 2.02/16 valves installed.

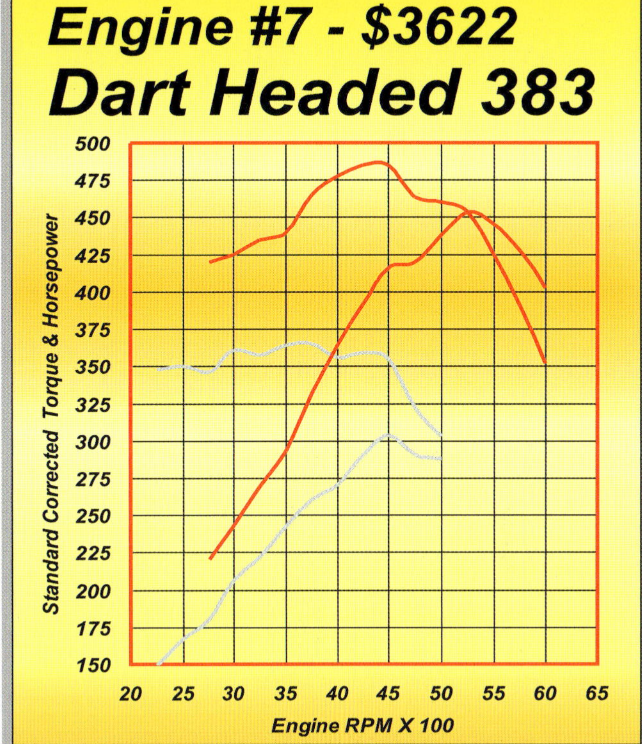

Intake: Edelbrock Performer RPM Air Gap two plane.

Carb: Vacuum secondary AED 750 series Holley with shaft and butterfly mods plus choke horn removed and blended out. Final flow: 815 cfm.

Air filter: K&N

Ignition: Performance Distributors HEI

Plugs/cables: Autolite/Accel

Headers: Walker 1⅝-inch.

Mufflers: Walker Dynomax 2½ inch.

Parts: $3004
Labor: $618
Total: $3622

This 383 unit was a really nice street driver. It ran well everywhere in the RPM range, making it docile in traffic but with enough punch to show well at the drag strip. It made strong torque right down to 1,000 rpm but the figures only start at 2,700 because the dyno would not load it down any lower. If you have a truck application in mind this could be what you want because it will tow a 5,000-pound load without a sweat.

Engine # 8: Mix-n-Match 383—Street or Strip

This motor was used as the basis of several magazine articles and reconfigured at each. What I intend to show here are builds with two different head, cam, and induction specs. You get to choose which way you want to go. This build started life with a clean 880 roller block of about 1993-ish vintage. This was bored to accept a set of flat-top KB pistons (KB 718) which, with the block decked to leave them 0.005-inch in the hole, delivered a CR of 10.5:1 with the 66-cc chambers in the heads. As for the heads, I used Dart Pro 1s here. For the first time around they were, with the exception of a little chamber work to bring them up to 66 cc from 64, as cast. These heads were the pre-Platinum design so you can expect better results from the current Dart Pro 1 Platinum heads.

The crank used was a Scat cast steel 3.75 stroker and was paired with a set of Scat Premium 7/16 stroker rods. This and a Professional Products crank damper completed the rotating assembly.

Because there are a lot of cubes under them, it is important that the heads selected flow well. The Darts chosen were the 200-cc intake port volume ones and were run stock (except for the small amount of chamber enlargement) in the first instance and with a basic porting job in the second.

The cam for the first build was a flat tappet, single-pattern COMP Xtreme Energy grind (profile number 5444). This 270-degree hydraulic profile, ground on a 106 LCA at 4-degrees advanced, lifts the valves to a little over 0.500 inch. The second cam used with the ported heads was a 5448 profile from the same series. This 290-degree profile, again on a 106 LCA 4-degrees advanced, lifted the valves to a little over 0.530 inch. In both cases a set of 1.6:1 COMP roller tip Magnum rockers were used.

Induction was via a used, and originally custom-built, mechanical secondary 750 Holley with the choke horn removed and blended in and the throttle spindles thinned down. The carb builder also installed stepped dog leg boosters so as to improve fuel atomization On my bench the carb flowed a solid 835 cfm. The first intake to be tested here was a Professional Products Crosswind two-plane. The second was an Edelbrock single-plane Super Victor race manifold. In both instances the runners were port matched to the heads and the plenum entry area tidied up as required. That involved minimal work on the two-plane, but quite a lot on the single plane. For the ignition, a Performance Distributors race spec HEI was used.

As can be seen from the output curves the street version of this 383 was an easy bolt-it-together deal to generate 488 ft-lbs of torque and 455 hp. Given the traction, that's enough grunt to put a typical 4,000-pound truck into the 13s on the quarter.

After doing just a basic porting job on the Dart heads, swapping out the intake, and installing the bigger COMP flat tappet hydraulic cam, the output climbed dramatically. Although torque below about 4,000 rpm dropped, peak torque went up slightly to 496 ft-lbs and peak horsepower to 540. In this form it won't be quite an ideal truck motor but it none-the-less can still be used in a stick-shift street vehicle without having low speed drivability problems.

Core: 350-ci 880 short-block from AAEQ. 2008 cost with shipping: $360.

Block: 880 roller cam 4-bolt block bored +0.030 over and decked to 9.005 inches.

Pistons: KB 718 forged flat-top 0.030 oversize.

Final CR: 10.6:1

Crankshaft: Scat 3.75 9000 series cast steel stroker crank.

Rods: Scat 6-inch Premium 7/16 stroker rods.

Oil Pump: Reworked stock.

Pan: Stock

Cam: Build number-1: COMP Cams Xtreme Energy Hydraulic flat tappet single-pattern 270 (Profile number 5444) on a 106 LCA in at 4 advance.

Build number-2: COMP Cams Xtreme Energy Hydraulic flat tappet single-pattern 290 (Profile number 5448) on a 106 LCA in at 4 advance.

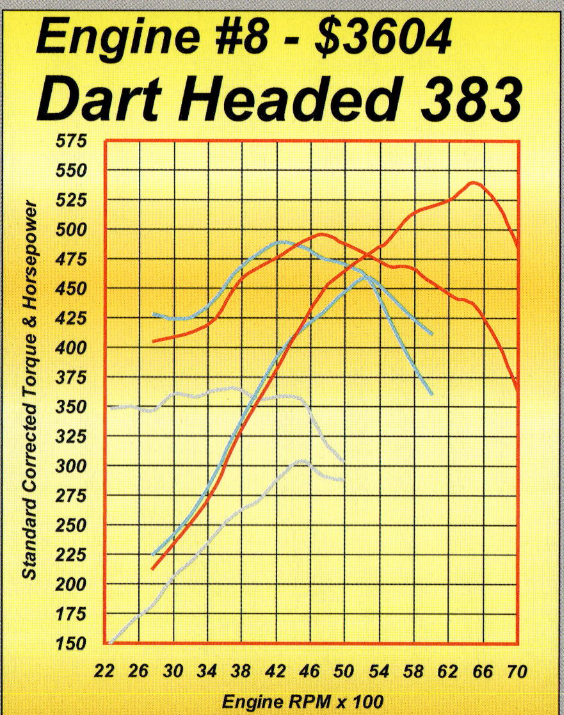

Timing set: COMP Cams adjustable.

Lifters: COMP 858-16 flat tappet hydraulic.

Valvetrain: COMP 987 springs. COMP Magnum roller tip rockers 1.6 intake and exhaust. COMP Magnum pushrods.

Heads: Build number-1 As cast Dart Pro 1
Build number-2 Basic ported Dart Pro 1

Intake: Build number-1 Professional Products Crosswind
Build number-2 Edelbrock Super Victor

Carb: AED 750 series Holley with shaft and butterfly mods plus choke horn removed and blended out. Final flow: 835 cfm.

Air filter: K&N

Ignition: Performance Distributors HEI

Plugs/cables: Autolite/Accel

Headers: Walker 1⅝-inch.

Mufflers: N/A

Parts: $3164
Labor: $440
Total: $3604

With this build you get a choice of favoring either the street or the strip as a primary application. Build number-1 (light blue curves) is all business for the street, while build number-2 (red curves), though street drivable, significantly favors the drag strip. Note that though there is a 70-hp difference in output the cost of either option is the same. About 20 hp is due to the cam—the other 50 due to the single plane Super Victor intake and head porting. The fact that build number-2 makes more horsepower is not the only contributing factor toward lower ETs at the strip. Because this build turns over 1,000 more useful rpm, it can be geared slightly lower in the first instance and it can hold a lower gear longer in the second instance.

Engine # 9: Mix-n-Match 408—Street or Strip

To some extent this build is similar to the number-8 build just covered. The differences here are that the 350 880 block has been clearanced to take a 4-inch-stroke crank. This requires the block be selected with a sonic tester. Rather than go traipsing around swap meets with my sonic tester and then going to all the fuss of building the short block, I took the easy way out and obtained a balanced and assembled short block (TWPE in Roanoke is a good source here). This featured off-the-shelf Mahle pistons adapted for use in this application (valve pockets modified), Scat Premium 7/16 stroker rods, and a Scat forged 4-inch-stroke crank. All this is good for a build with a redline at 7,400 rpm, which is more than enough for a 600-hp-spec motor. For the first round this bottom end assembly was used with a COMP Xtreme Energy hydraulic roller cam, ported Dart Pro 1 Platinum heads (200-cc runners), and an Edelbrock Performer RPM Air Gap two-plane intake. The carb was an AED Holley 850, which checked out at 930 cfm on our bench.

Picking the right cam for this engine was an important issue. During porting I had stepped up the intake valve for the Dart heads from 2.02 to 2.08. In terms of cam LCA, this meant that, although the cubes had gone up, I could still stay with a 106 LCA for the cam because the valve size had also gone up over what is typically used (2.02/2.05) in a 383-ci motor. The cam spec then was a number 3194 on the intake and a 3196 on the exhaust. These lobes (230/236 degrees at 0.050 lift) were on a 106 LCA at 4 advance. With a 1.6/1.5 rocker combo this resulted in 0.622 intake lift and 0.585 exhaust lift.

The blue curves show the results from the Dart-headed spec. As can be seen, there is plenty of torque on hand. This engine would pull right down to a little over idle but this does not show on the graph because the torque was too high for the dyno to load it down to these lower RPM figures.

Moving on to the AFR headed version of this engine, we see the type of results that can be had by AFR's entry-level CNC 195 street Eliminator heads. Although the ports might look to be a little on the small side for a 408-inch engine, they none-the-less delivered a strong top end as well as the expected big torque numbers. Starting at the point of air entry this spec of our 408 utilized the same AED carb but this time around it was mounted on a port- and plenum-matched Super Victor intake. The 2.05/1.6-inch valves of the AFR Eliminators were operated by a set of COMP 1.65 intake 1.5 exhaust rockers. These delivered, after lash, 0.622 lift on the intake and 0.570 on the exhaust. The cam was a solid COMP Xtreme Energy street roller profile number 4876 on the intake and number 4877 on the exhaust. These grinds give 254/260 at 0.050 lift with an advertised duration of 292/298. The lobe centerline angle was 105 at 4-degrees advance.

Between the race intake manifold and a near race cam, this combination (red curves) lost about 100 ft-lbs down low. But as the RPM rose to about 4500, everything started to chime in and resulted in a peak torque of 530 ft-lbs and 595 hp—all on pump gas!

If you compare this 408 with our number-1 build, that's the light gray curves on the graph, you will see that we are up on torque by 50 percent and power has all but doubled.

Core: 350-ci 880 short-block fully prepared with stroker crank installed. 2008 cost with shipping: $1,600.

Block: 880 roller cam 4-bolt block bored +0.030, decked to 9.005 inches and clearanced for a 4-inch stroker crank.

Pistons: Modified Mahle pistons to suit application 0.030 oversize.

Final CR: 10.5:1

Crankshaft: Scat 4.000 stroke forged two-piece rear main seal (requires rear main seal adaptor kit).

Rods: Scat 6-inch Premium 7/16 stroker rods.

Oil Pump: Reworked stock.

Pan: Moroso

Cam: Build number-1- COMP Cams Xtreme Energy Hydraulic roller 230/236 at 0.050 (Profile number 3194 In. 33196 Ex.) on a 106 LCA in at 4 advance. Build number-2 - COMP Cams Xtreme Energy solid street roller 254/260 at 0.050 (Profile number 4876 In. and number 4877 Ex.) on a 105 LCA in at 4 advance.

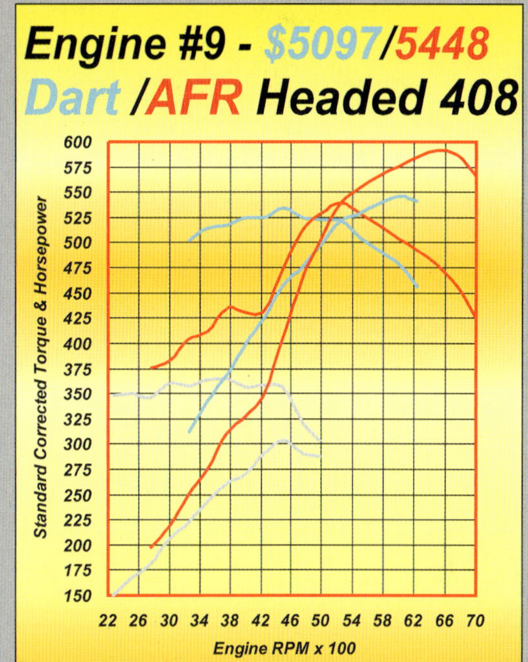

Timing set: COMP Cams adjustable.

Lifters: Build number-1 COMP 875-16 hydraulic.
Build number-2 COMP 888-16 solid.

Valvetrain: Ordered heads come with appropriate springs for whatever cam is used. COMP aluminum roller rockers 1.6 intake and 1.5 exhaust. COMP Magnum pushrods.

Heads: Build number-1 Ported Dart Pro 1
Build number-2 CNC ported AFR straight plug street Eliminator 195

Intake: Build number-1 Edelbrock Performer RPM Air Gap two plane. Build number-2 Edelbrock Super Victor single plane race.

Carb: AED 850 series Holley with shaft and butterfly mods plus choke horn removed and blended out. Final flow: 930 cfm.

Air filter: K&N

Ignition: Performance Distributors HEI.

Plugs/cables: Autolite/Accel

Headers: Walker 1¾-inch.

Mufflers: N/A

Parts: Build number-1: $5097
Build number-2: $5448
Labor: N/A
Total: As per above

Purchasing the 408 short block already done meant that there were no labor costs involved with this build other than re-cutting the valveseats of the ported Dart heads. Other than this, it was all assembly work right from the start. Build number-1 (light blue curves) delivered what it takes to make a true street motor into outstanding. As for the AFR Street Eliminator-headed version of this engine, we see that it came really close to cracking the 600-hp mark. One of the targets for this engine was that even in "race" form it had to run pump gas. All these curves are just that—on pump gas. Had we opted to put the CR up on the big-cammed AFR-headed engine it would have sailed by the 600-hp mark with ease as well as made more torque—especially down below the peak torque RPM.

Engine # 10: 434 Torque Monster

This engine started life as a bare 400 block that sonic-tested out pretty good and was in need of a re-bore. Unlike a lot of 400 blocks this one still had a stock-size set of bores. They were in bad shape and there was some doubt as to whether the bores would clean up at 30 over. As it happened they did, but only just. Originally a 2-bolt block one of the first jobs was to install a set of PRW angle-bolt 4-bolt mains caps. These angle bolt caps are stronger than the vertical-bolt 4-bolt factory caps so that's the way to go if you find a good 2-bolt block. By the time the mains were align bored and the rest of the block machining done, I was into the block for about $800 plus the cost of the caps.

For a crank I used Scats lowest-priced regular forged 4-inch-stroke crank and a set of 6-inch Premium 7/16 rods. Pistons were flat-top KBs (P/N KB849). Rounding off the bottom end, a Moroso oil pan for 4-inch strokers was used. This is important because these long stroke cranks can gather up oil and throw away 20 to 30 hp in the process.

Carb for this build was a maxed-out AED 950-style Holley that flowed just short of 1,000 cfm. The manifold was a port matched Professional Products Crosswind that had some basic porting done on it in the areas most accessible. The AFR heads were just as they came out of the box. With the KB flat-top pistons and the 75-cc head chambers the CR was a healthy 10.7:1. That's a little more than the 10.5:1 that I normally target for a pump gas motor, but with a 180-degree thermostat no problems were seen with a good brand of premium pump fuel.

The cam for this 434 was a COMP Xtreme hydraulic roller with a 288/293 advertised duration. This at 0.050 was 236/242. The profiles (numbers 3196 and 3197) were ground on a 104 lobe centerline angle at 4 advance. Rockers used were 1.6:1 all around. This tight LCA and the fact that it is timed in to 100 degrees ATDC means we are using up valve-to-piston clearance. So like any build with a big-ish cam in well advanced, the clearance needs to be checked. When the valvetrain dynamics are basically an unknown, it's best to have at least 0.100 clearance between the valves and the pistons.

For ignition, a Performance Distributors HEI was used.

The resulting output from this relatively straightforward build made for the ideal street power plant. It had tire-shredding torque everywhere in the RPM range yet would cruise perfectly at 1,000 rpm in high gear. The best analogy I can think of here is that it was like having the latest LS injected 427 Corvette engine hopped up about 40 hp, but at less than half the price!

Core: 400-ci short block from local parts salvage yard. 2008 cost: $350.

Block: 2-bolt 400 block converted to angle cap (PRW) 4-bolt mains bored +0.030, decked to 9.000 inches and clearanced for a 4-inch stroker crank.

Pistons: KB 849 flat-top 0.030 oversize with valve pockets cut deeper.

Final CR: 10.2:1

Crankshaft: Scat 4.000 stroke forged two-piece rear main seal (requires rear main seal for align bored mains).

Rods: Scat 6-inch Premium 7/16 stroker rods.

Oil Pump: Reworked stock.

Pan: Moroso

Cam: COMP Cams Xtreme Energy Hydraulic roller 236/242 at 0.050 (Profile number 3196 intake number 3197 exhaust) on a 104 LCA in at 4 advance.

Timing set: COMP Cams adjustable.

Lifters: COMP 875-16 hydraulic rollers.

Valvetrain: Ordered heads come with appropriate springs for whatever cam is used. COMP aluminum roller rockers 1.6 intake and 1.5 exhaust. COMP Magnum pushrods.

Heads: CNC ported AFR 75-cc chamber straight-plug Street Eliminator 195.

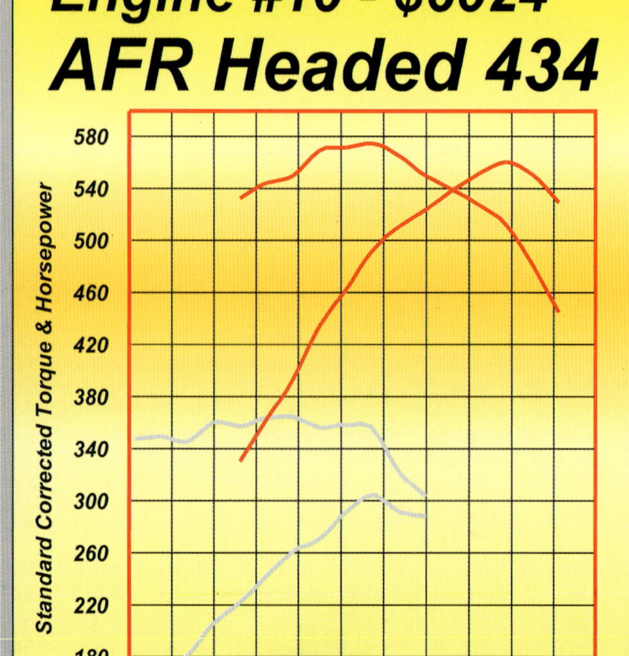

Intake: Professional Products Crosswind air-gap-style two-plane port, matched and blended about 2 inches into the runners from the manifold faces.

Carb: AED 950 series Holley with shaft and butterfly mods plus booster and main venturi mods for a final flow of about 980 cfm.

Air filter: K&N

Ignition: Performance Distributors HEI.

Plugs/cables: Autolite/Accel

Headers: Walker 1¾-inch.

Mufflers: N/A

Parts: $5144
Labor: $880
Total: $6024

This build is a prime example of the old adage that there is no substitute for cubic inches. This build has it all—big torque numbers everywhere as well as big horsepower. On the street it drove like a big-block, but the car handled like a small-block machine with none of that front-heavy feel about it. As for mileage, this is good so long as this motor's ability to lug right down to near idle RPM is utilized. A tall high gear or an overdrive unit will allow a 70-mph cruise with 1,500 to 1,800 rpm on the tach while still retaining strong get-up-and-go when needed, without a down shift.